FORMABLE HSLA AND DUAL-PHASE STEELS

Proceedings of a symposium sponsored by the TMS-AIME Ferrous Metallurgy Committee at the Fall Meeting held in Chicago, Illinois, October 26, 1977.

Edited by
A. T. DAVENPORT
Republic Steel Research Center
Cleveland, Ohio

A Publication of

The Metallurgical Society of AIME

A publication of The Metallurgical Society of AIME

Printed in the United States of America. Library of Congress Card Catalogue Number 79-52167.
ISBN Number 0-89520-351-0.

FOREWORD

The papers in this volume were selected from a symposium titled "Modern Developments in HSLA Formable Steels" which was held during the Fall Meeting of The Metallurgical Society of AIME in Chicago, October 1977, under the sponsorship of the Ferrous Metallurgy Committee. The purpose of the symposium was to bring up to date the progress made in improving the formability of HSLA steels, with the emphasis on the metallurgical rather than the engineering aspects. These steels have undergone substantial development in the last decade, the prime target being the automotive industry, where the weight-saving potential of HSLA steels is of tremendous interest in the present drive for fuel economy. Unfortunately, the relatively poor formability of HSLA steels in comparison to mild steel has been a major factor restricting their widespread application. These proceedings highlight a major breakthrough in the formability of HSLA steels, namely, the dual-phase ferrite martensite steels. This phenomenon has all the hallmarks of being one of the most important technological developments of the decade and one which could play a key role in widening the application of HSLA steels.

A word on the discussion. The published discussion is that submitted in writing, and is extremely limited considering the active and extensive oral discussion at the symposium. Unfortunately, the attempts to record these oral discussions for publication were unsuccessful due to poor recording quality.

Thanks are due to the members of the Ferrous Metallurgy Committee for their support, and in particular to Professor J. D. Embury of McMaster University, Mr. B. Rothwell of the Noranda Research Centre, and Dr. P. R. Mould of United States Steel Corporation for their assistance in organizing and chairing the symposium.

A. T. Davenport
Republic Steel Corporation
Independence, Ohio

TABLE OF CONTENTS

SESSION II

Chairmen: J. D. Embury, McMaster University
B. Rothwell, Noranda Research Centre

RELATIONSHIP BETWEEN STEEL MICROSTRUCTURE AND FORMABILITY

M. S. Rashid
Metallurgy Department
General Motors Research Laboratories
Warren, Michigan 48090

The strain hardening characteristics and microstructural deformation behavior of the SAE 980X, 950X and plain carbon steels, having ferrite-carbide microstructures, and specimens with ferrite-martensite or dual phase microstructures which were produced by heat treating the SAE 980X steel were investigated. The yield and tensile strengths of all tested heat treated specimens were similar, but total elongation (e_t), which is directly related to formability varied from 23.4 to 31.8% depending on heat treatment cooling rate, an e_t of 20% being typical in the SAE 980X steel. Like most steels, the ferrite-carbide specimens exhibited parabolic strain hardening behavior; however, those with ferrite-martensite microstructures did not. All the heat treated specimens had significantly higher strain hardening coefficients than the SAE 980X steel. Plastic deformation was found to occur in the ferrite-carbide steels by deformation of the ferrite matrix, while in the heat treated steels with e_t tending towards the upper end of the range indicated, both the ferrite and martensite phases underwent deformation. It was concluded that a higher strain hardening coefficient of the ferrite and the presence of a deformable martensite phase were the major factors responsible for the high e_t values.

INTRODUCTION

Automotive component weight can often be reduced by substituting plain carbon steel with high strength steels, such as SAE 950X or 980X high strength low alloy (HSLA) steels. These steels have the same basic composition as plain carbon steel with minor additions of V, Cb or Ti. They are thermomechanically processed and are considerably stronger, but are less formable than plain carbon steel. However, it was recently shown (1) that the formability of vanadium strengthened SAE 980X steel of certain compositions, could be improved significantly (Fig. 1) by heating the steel above its eutectoid temperature for a few minutes and cooling to room temperature.

The microstructure of the SAE 980X steel consists of a ferrite matrix with grain boundary iron carbides (Fig. 2), a small volume fraction of pearlite and decomposed pearlite, and a dense distribution of fine vanadium carbonitride (VCN) strengthening precipitates. The heat treated steel microstructure (2) consists of a ferrite matrix with 10-20 volume percent of a second phase, tentatively identified as martensite (Fig. 3), which appears to have replaced the uniform distribution of iron carbides (Fig. 2), and redistributed VCN precipitates. Other transformation products or retained austenite may also be present. Steels with ferrite-martensite microstructures have previously been referred to as "dual phase" steels [3].

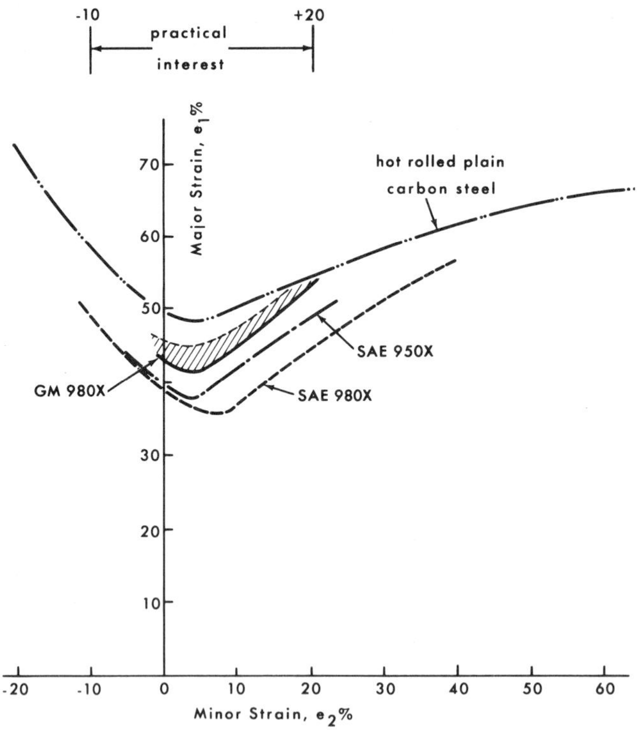

Fig. 1 - Range of the forming limit diagrams (FLD) of GM 980X compared with that of plain carbon steel and SAE 950X and 980X steels. The GM 980X was obtained by heat treating a SAE 980X steel (Ref. 1).

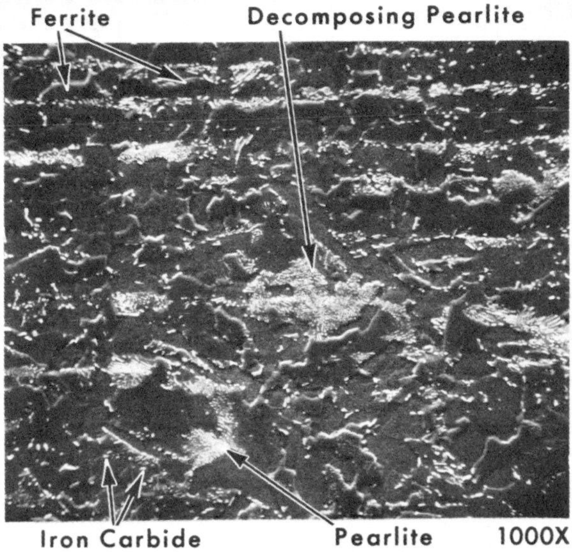

Fig. 2 - Scanning electron micrograph of the SAE 980X steel showing a matrix of small grained ferrite, grain boundary iron carbides, and some pearlite and decomposed pearlite (Ref. 2).

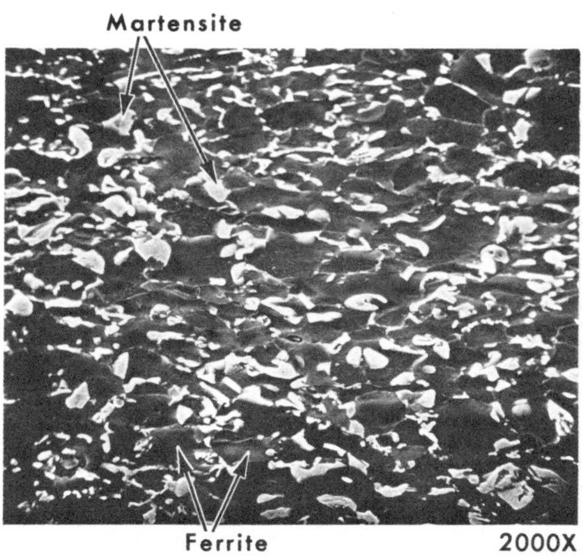

Fig. 3 - Scanning electron micrograph of GM 980X showing a matrix of ferrite with islands of martensite which appear to have replaced the uniform distribution of iron carbide particles in Fig. 2 (Ref. 2).

Nominal stress-strain curves for the SAE 980X steel before and after heat treatment are shown in Fig. 4. Data for SAE 950X and plain carbon steels which like the SAE 980X steel also have a ferrite-carbide microstructure are shown for comparison. Typically, a SAE 980X steel has a yield strength, YS, of 550 MPa (80,000 psi), an ultimate tensile strength, UTS, of 650 MPa (95,000 psi), a total elongation, e_t, of 20% in a 50.8 mm (2 in.) gage length ASTM E8 standard specimen, and a strain hardening coefficient, or n value, about 0.12. The yield point elongation, ype, shown in the figure may vary from 0 to 3%. The heat treated SAE 980X steel [1,2] had a much lower YS of 350 MPa (50,000 psi), no ype, work hardened to a YS of 550 MPa (80,000 psi) with 2 to 3% strain, a UTS equivalent to that of SAE 980X steels, a much larger e_t, typically 30%, and a strain hardening coefficient more than double that of SAE 980X steels. Irrespective of chemistry or processing parameter differences, steels which have these mechanical properties are designated GM 980X.

Mechanisms responsible for the shape of the GM 980X σ-ε curve at low strains have been fairly well explained [2]. The martensite transformation generates an abundance of new free dislocations in the parent ferrite matrix. Movement of the unobstructed dislocations results in

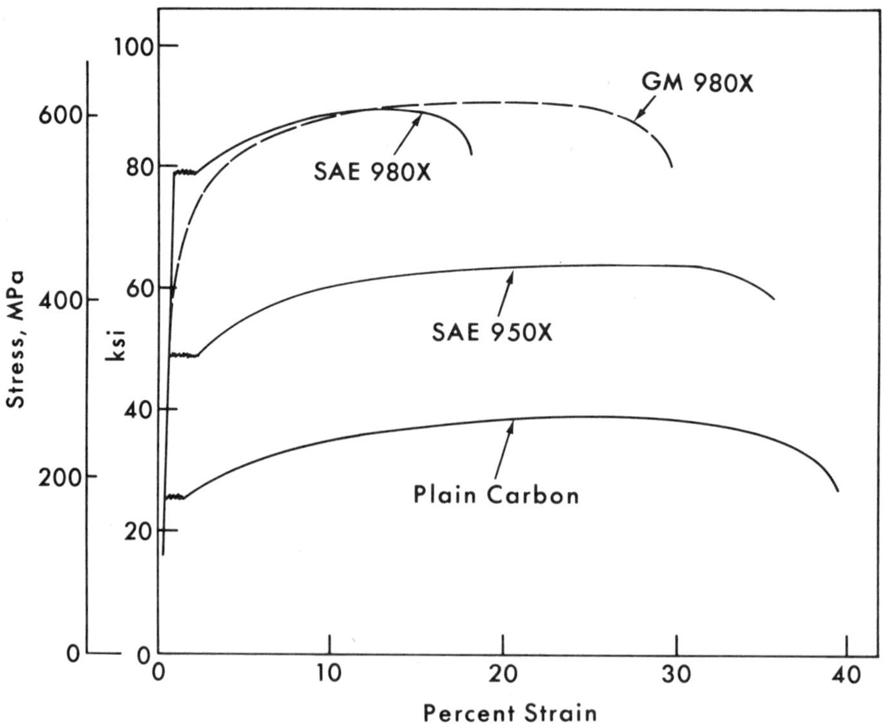

Fig. 4 - Nominal stress-strain curves for plain carbon, SAE 980X, SAE 950X and GM 980X steels. Strain indicated is for 50.8 mm (2 in) gage length specimens.

the absence of ype and the observed low YS. Interaction of the increased number of dislocations with each other and with the finely dispersed secondary phases results in the high strain hardening coefficient. Mechanisms responsible for the large e_t accompanying the high UTS and the observed superior formability have not been thoroughly examined, however. This investigation takes a closer look into the causes for the superior GM 980X formability via a study of the strain hardening characteristics and microstructural deformation behavior of two groups of steels. The first group consists of steels with a ferrite-carbide microstructure, namely, the SAE 980X, SAE 950X and plain carbon steels. The second group consists of steels with ferrite-martensite or "dual phase" microstructures which were produced by heating the SAE 980X steel for four minutes at 788°C (1450°F) and cooling to room temperature at different rates.

Strain hardening behavior can be evaluated by determining the n value of the steel. Physically, n gives a measure of the ability of a metal to distribute strain [4]. The higher the n value, the more uniform the strain distribution and, hence, the greater the resistance of the metal to necking. A high n value accompanied by a large e_t is indicative of superior formability. The n value is quite easily obtained from tension test data [5]. The true stress-true strain curve for most metals, especially steel, can be approximated by the equation:

$$\sigma = K\varepsilon_p^n \tag{1}$$

or,

$$\log \sigma = \log K + n \log \varepsilon_p \tag{2}$$

where σ = true stress

ε_p = true plastic strain

K = strength coefficient; σ at $\varepsilon_p = 1.0$

n = strain hardening coefficient.

When Eq. 1 is satisfied, a log σ-log ε_p plot will be a straight line with n being the slope. Frequent deviations from Eq. 1 are observed, however. One common type of deviation is for the log σ-log ε_p plot to result in two straight lines of different slopes, while in other cases a curve with a continuously changing slope is obtained. In the latter case, an incremental value of n, n_i, must be calculated for small increments of ε_p to evaluate the strain hardening behavior of the metal.

Localized necking, or plastic instability, begins in a tension test at maximum load, P_{max}, i.e., when dP/de=0. Whether Eq. 1 is satisfied or not, ε_p at this instability condition will numerically equal n_i and is usually referred to [5] as the uniform strain, ε_u. When Eq. 1 is satisfied, n_i and n are the same and each equals ε_u. This paper describes the variations of n_i and microstructural deformation with increasing strain of each steel evaluated. It also attempts to show a correlation between the two parameters and their relationship to steel formability.

Experimental Procedure

Standard size (ASTM E8) tensile specimens were machined from flat steel sheet with their axes parallel to the steel rolling direction. In order to obtain dual phase microstructures, tensile specimens were heated in a neutral salt bath and then cooled to room temperature with forced air at cooling rates ranging from 5° to 14°C/sec (9° to 26°F/sec). The cooling rates were designated A, B, C and D; A being the slowest and D the fastest. The tension tests were conducted in a Wiedemann Baldwin Universal Testing machine at a constant crosshead velocity of 5.0 mm/min (0.2 in/min). Strain was measured with a Satec dual range extensometer over a 50.8 mm (2 in) gage length. Load-strain data were automatically recorded on a strip chart recorder. The X-Y coordinates were measured off the chart for strain increments of 0.00625 or 0.0125 and a computer program was used to analyze the data. Calculation of n_i is explained in the Appendix. True stress, ε_p, and n_i were calculated for each strain increment and log σ vs log ε_p, and n_i vs ε_p curves were plotted.

The microstructural deformation characteristics of the tested tensile specimens were examined in the rolling direction through the sheet thickness with a Cambridge Scanning Electron Microscope (SEM). The specimens were mounted in Bakelite and prepared for SEM examination as described in Ref. 6.

RESULTS

Strain Hardening

Mechanical properties of the as-received SAE 980X, 950X and plain carbon (ferrite-carbide) steels and their chemistries are listed in Tables I and II, respectively. Log σ vs log ε_p curves for the steels are plotted in Fig. 5. Each set of data can be represented by a straight line with a constant slope. The slope or n value for each steel was equal to ε_u measured in the tension test (Table I). As expected, plain carbon steel had the highest n value of the three steels and the SAE 980X the lowest.

Mechanical property data for the heat treated SAE 980X steel (ferrite-martensite) specimens are listed in Table III and plotted in Fig. 6. Irrespective of cooling rate, UTS was ~685 MPa (99,000 psi) and similar to that of the SAE 980X steel, and YS decreased from 550 MPa (80,000 psi) in the SAE 980X steel to ~310 MPa (45,000 psi) in the heat treated specimens. This was in agreement with earlier results [1]. However, e_t changed significantly with increasing cooling rate. A maximum e_t of 31.8% was observed for cooling rate B (6°C or 11°F/sec), and e_t was between 27 and 28% for cooling rates A and C. For D, the fastest cooling rate evaluated (14°C or 26°F/sec), e_t was lower still and only slightly higher than that for the SAE 980X steel. Specimens with cooling rates A, B and C, therefore, had mechanical properties which met the GM 980X specification (Table IV) while the specimen with cooling rate D did not.

Table I. - Mechanical Properties of the As-Received Steels.

Steel	Yield Strength MPa	Yield Strength psi	Ultimate Tensile Strength MPa	Ultimate Tensile Strength psi	Total Elongation in 50.8 mm (2 in) GL	n Value	True Uniform Strain, ε_u
SAE 980X	548	79500	682	99000	0.22	0.12	0.12
SAE 950X	380	55200	462	67000	0.27	0.14	0.14
Plain Carbon	232	33600	284	41100	0.38	0.20	0.20

Table II. - Chemical Composition* of the Steels Evaluated.

	Weight %									
Steel	C	V	Mo	Si	Mn	Al	P	S	N	O
SAE 980X	0.12	0.11	0.008	0.51	1.46	0.11	0.008	0.009	0.019	0.002
SAE 950X	0.09	0.05	0.011	0.12	0.83	na	na	na	0.011	0.004
Plain Carbon	0.05	na	na	0.02	0.32	na	na	na	na	na

*Analyzed by GMR Analytical Chemistry Department

na - not analyzed

Log σ-log ε_p curves for the heat treated SAE 980X steels are plotted in Fig. 7. The curve for the as-received SAE 980X steel is replotted for comparison. While the SAE 980X data can be represented by a straight line with a constant 0.12 slope, only a portion of the data for each heat treated specimen may be approximated as a straight line with a slope or n value of 0.33, 0.35, 0.33 and 0.25 for cooling rates A, B, C and D, respectively. These n values were much higher than ε_u measured in the tension test (Table III), however. Data for the three specimens with e_t > 27% grouped together, while those for cooling rate D which had the poorest e_t of the heat-treated specimens tended toward the SAE 980X curve.

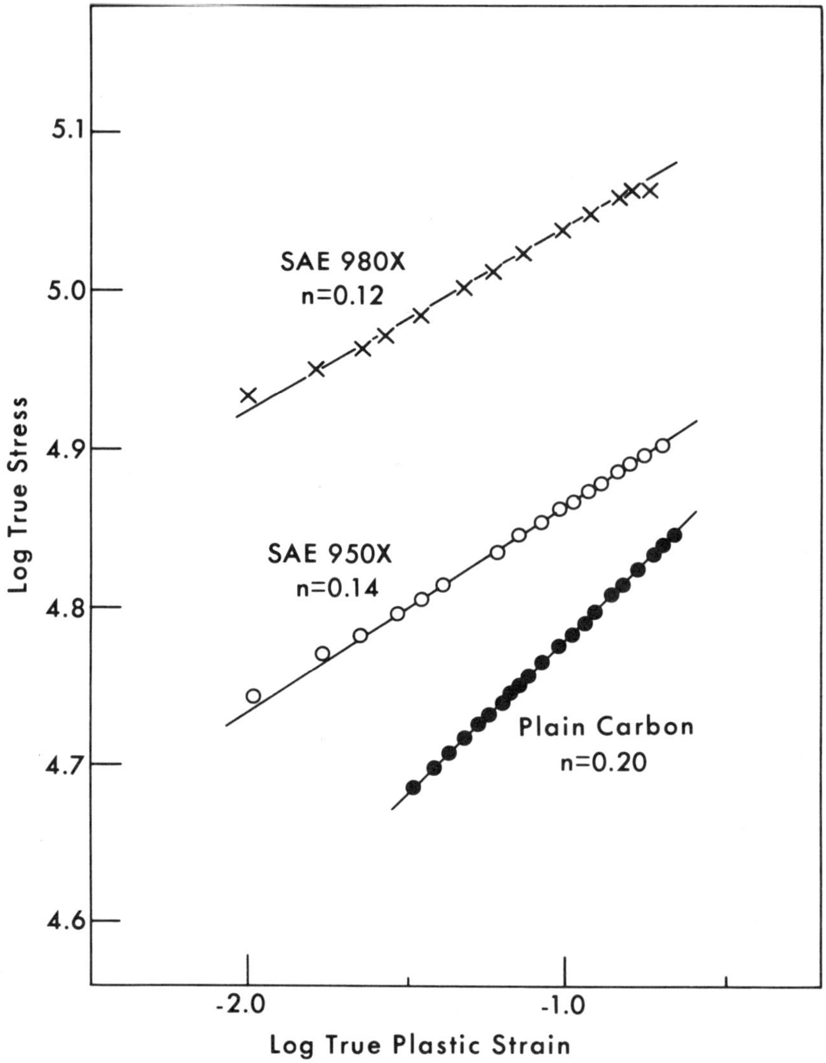

Fig. 5 - Log σ-log ε_p curves for the as-received steels.

Variations of n_i for small increments of strain for the three as-received steels are plotted in Fig. 8. The line for $n_i = \varepsilon_u$ which represents onset of necking is also shown here. In all three steels, n_i increased from practically zero at very small strains to the constant n value obtained in the log σ-log ε_p plots (Fig. 5) at a strain of about 0.05. Thereafter, it remained practically constant at the full curve n value until the onset of necking.

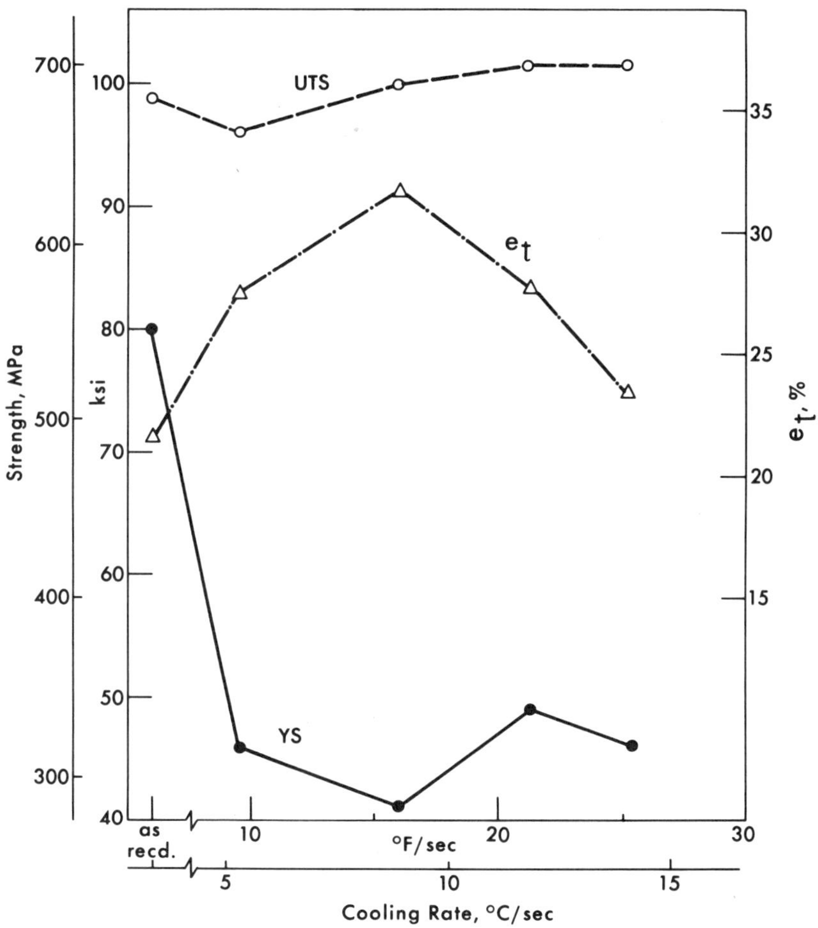

Fig. 6 - Effect of Cooling Rate, after heating for four minutes at 788°C (1450°F), on the mechanical properties of the SAE 980X steel.

Table III. - Mechanical Property Data for Heat Treated Specimens Tested.

Cooling Rate			Yield Strength		Ultimate Tensile Strength		Total Elongation in 50.8 mm (2 in)	Uniform Elongation	True Uniform Strain	n_i at
Designation	°C/sec	°F/sec	MPa	ksi	MPa	ksi	gage length, e_t %	e_u	ε_u	P_{max}
-	-	-	548	79.5	683	99.0	21.7	0.13	0.12	0.12
A	5.2	9.4	316	45.8	661	95.9	27.5	0.22	0.20	0.20
B	8.9	16.0	282	40.9	689	99.9	31.8	0.26	0.22	0.21
C	11.8	21.2	339	49.1	700	101.5	27.8	0.22	0.20	0.20
D	14.1	25.3	317	46.0	701	101.7	23.4	0.18	0.17	0.16

Table IV. - GM 980X Mechanical Properties.

	Specified Minimum or Maximum	Typical
Yield Strength, 0.2% offset	380 MPa (55 ksi) max	350 MPa (50 ksi)
Yield Strength, 3.0% offset	480 MPa (70 ksi) min	550 MPa (80 ksi)
Tensile Strength	620 MPa (90 ksi) min	650 MPa (95 ksi)
Elongation in 50.8 mm gage length	27% min	30%

In the heat treated specimens, n_i was significantly higher than that for the as-received HSLA steels over the entire range of strain (Fig. 9). It increased rapidly at low strains and then decreased rapidly from a very high value to the n value obtained from the linear portion of the log σ-log ε_p plots (Fig. 7) at an $\varepsilon_p \simeq 0.03$, and then further decreased at a slower rate with increasing strain. At the onset of necking, n_i in each specimen was approximately equal to ε_u measured in the tension test (Table III). Again data for cooling rates A, B and C are grouped together, but differences are apparent in specimen strain hardening behavior. As observed previously, data for cooling rate D do not fall in the same group and tend towards that for the SAE 980X steel.

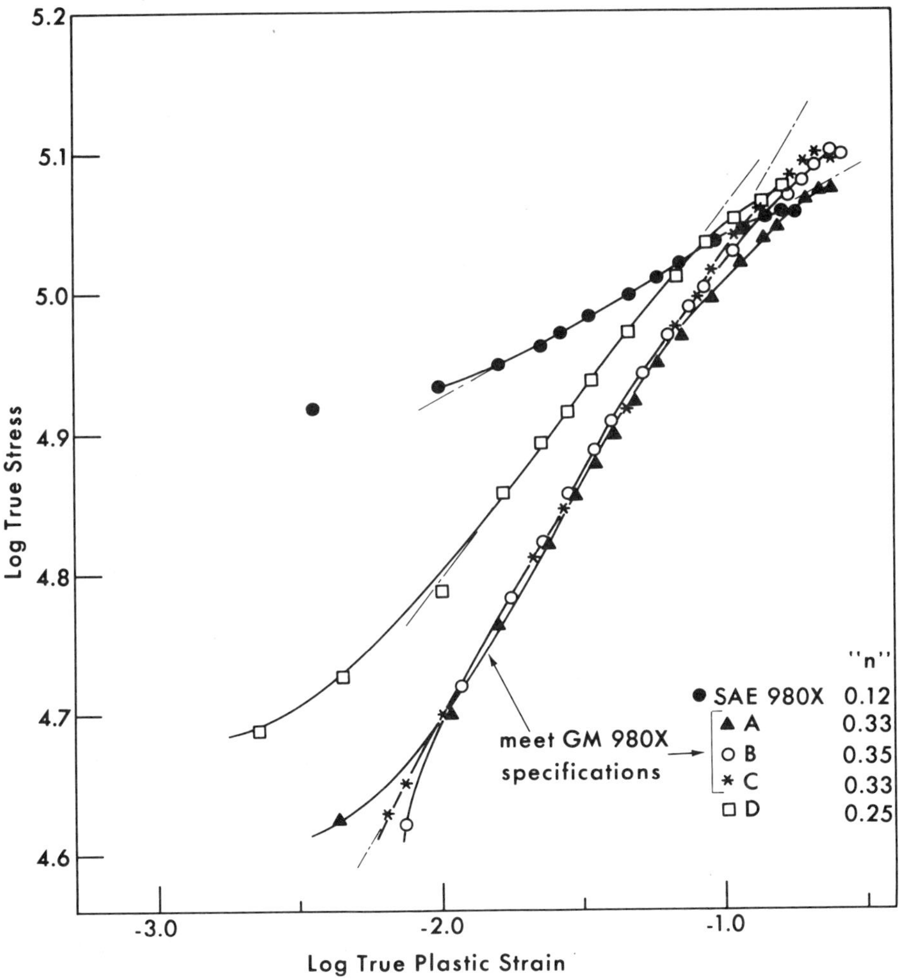

Fig. 7 - Log σ-log ε_p strain plots for the as-received and heat treated SAE 980X specimens. The dotted straight lines indicate the average full curve "n" values listed.

Specimens with cooling rates A and C had the same e_t, ε_u and n value or slope of the linear portion of the log σ-log ε_p plots. However, variation of n_i with ε_p was different in the two specimens (Fig. 9). For cooling rate C, n_i remained constant at 0.33 from ε_p = 0.03 to 0.09 and then decreased gradually with increasing strain to 0.20 at the onset of necking. For cooling rate A, however, the decrease in n_i occurred for ε_p > 0.05.

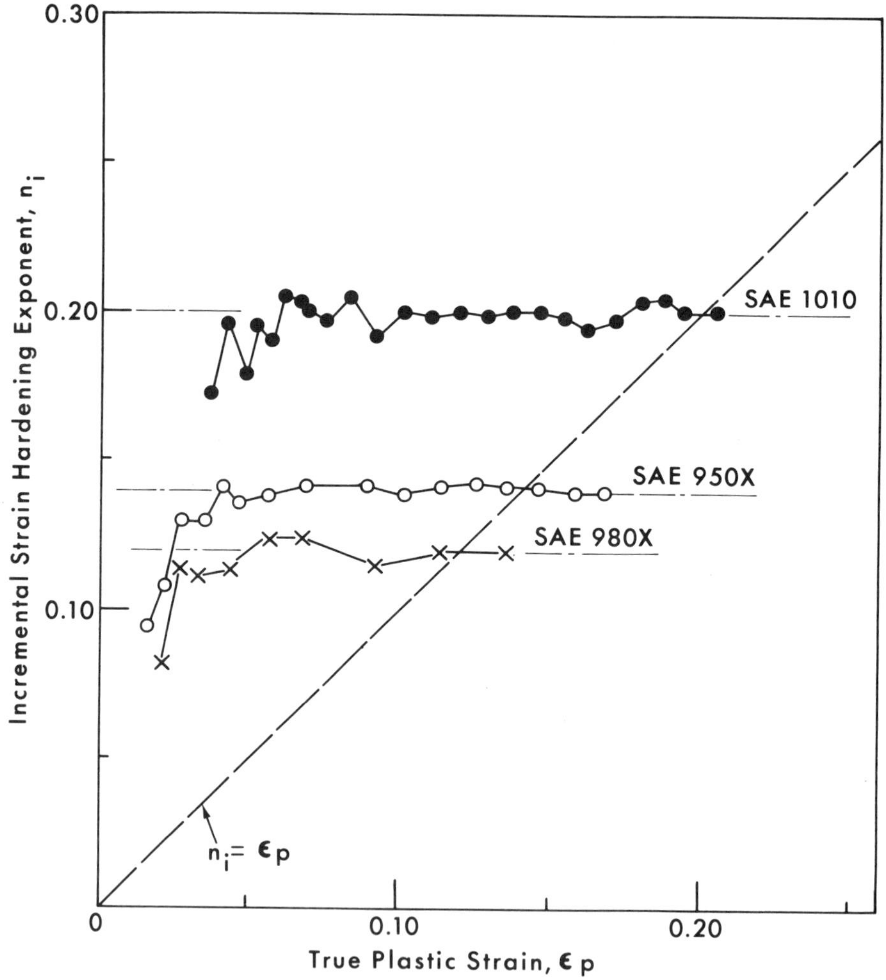

Fig. 8 - Variations of n_i with ε_p in the as-received steels. The onset of necking occurs at the intersection of dotted line with the curves. The three horizontal lines indicate full curve n values for each steel from Fig. 5.

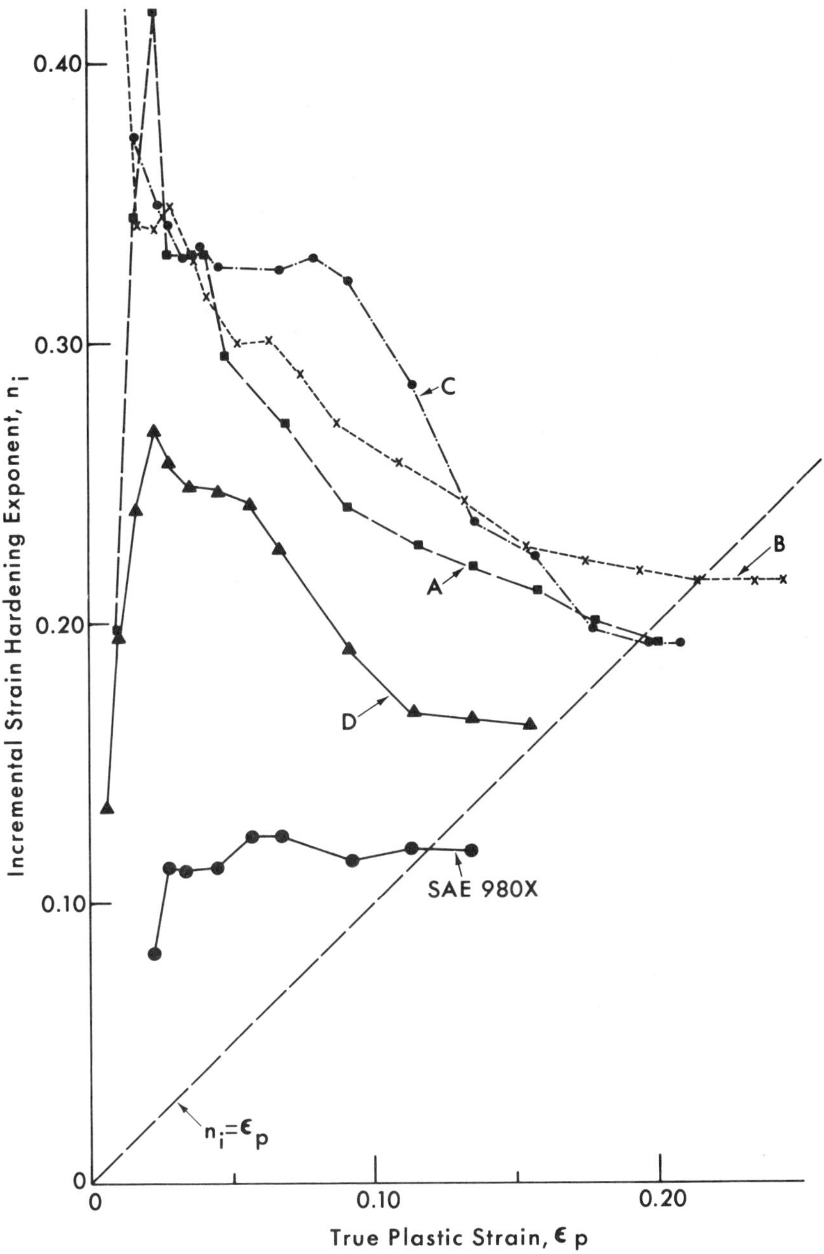

Fig. 9 - Variation of n_i with ε_p in as-received and heat treated SAE 980X specimens. The onset of necking occurs at the intersection of dotted line with the curves.

Microstructural Deformation

The areas examined by SEM on tension specimens are shown in Fig. 10. Following deformation, the parent ferrite grains were elongated in all steels (Figs. 11a, 12a, 13a, 14a). Because of their random shape and distribution, the change in the shape of the iron carbide particles in the as-received steels and the martensite islands in the heat treated steels, after deformation could not be established with certainty in areas of low strain. However, much was learned about the steel deformation behavior by examining the microstructures within highly strained areas, such as, the necked region. At location B in the as-received steels, the parent ferrite grains were severely elongated (Figs. 11b, 12b, 13b), the iron carbide particles were fractured and voids were visible in the ferrite matrix around the iron carbides. Deformation of the carbides was not immediately obvious. Some alignment of the carbides was evident, however. The fractured iron carbide particles were identified by their sharp edges. Their number and the number of accompanying voids (Figs. 11c, 12c, 13c) increased with decreasing distance from the fracture (A in Fig. 10).

As observed in the preceding as-received steels, the ferrite grains within the necked regions (area B) of the heat treated steels were also severely elongated (Fig. 14b). In addition, it was quite evident that the martensite islands in specimens with the three slower cooling rates (Table III) were also elongated and had apparently undergone deformation. Also, a few voids were present in the ferrite around the deformed martensite islands. Similar deformation of the martensite phase was less evident (Fig. 15a, b), however, in the heat treated specimen with the fastest cooling rate (D, Table III) and the volume fraction of martensite appeared to be somewhat higher. Moving closer to the fracture (A in Fig. 10) some of the martensite islands appeared to be ruptured and the number of voids were found to increase (Figs. 14c and 15c).

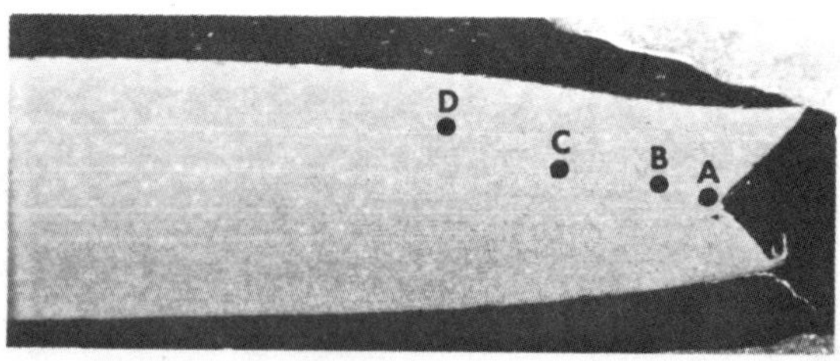

Fig. 10 - Locations examined by SEM on a tested tensile bar (20X).

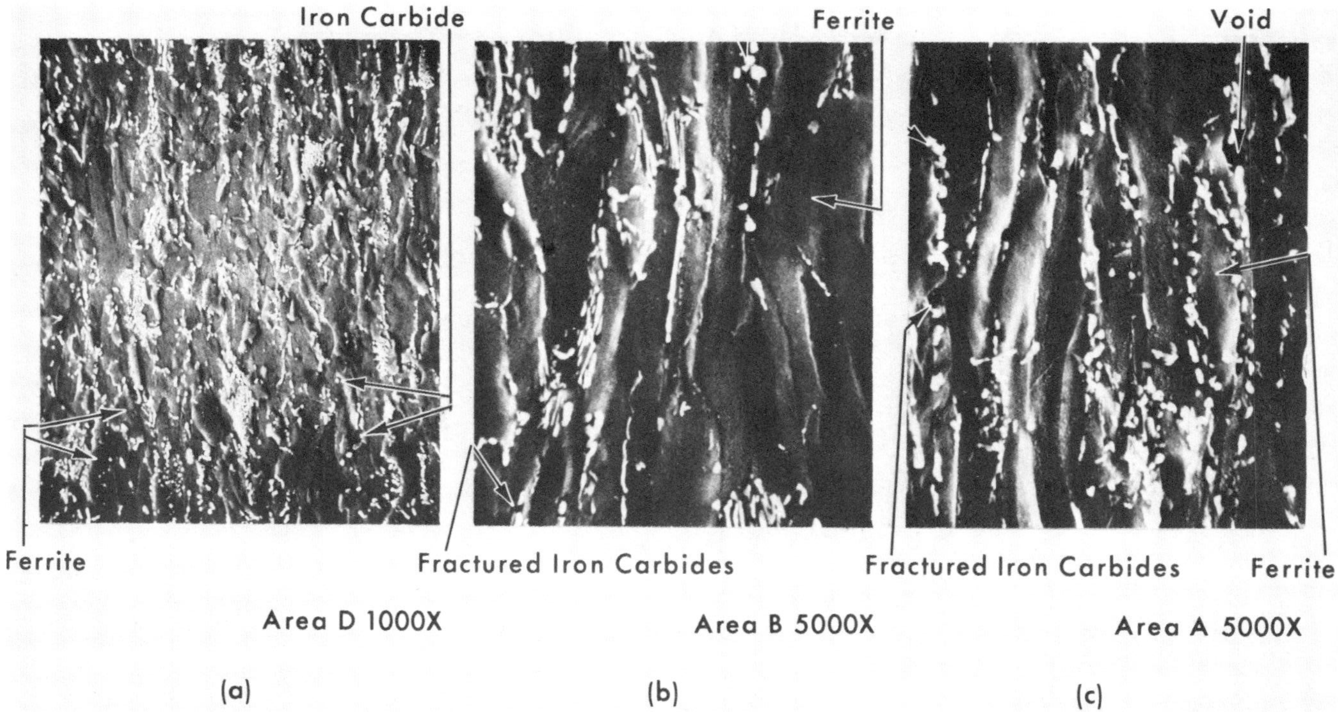

Fig. 11 - Scanning Electron Micrographs of deformed SAE 980X steel. (a) Elongation of the parent ferrite phase is evident. (b) The ferrite grains are grossly elongated and the iron carbides are aligned. Some particles are fractured and voids are formed at the ferrite-iron carbide interface. (c) A larger number of voids and fractured carbides are present here than in (b).

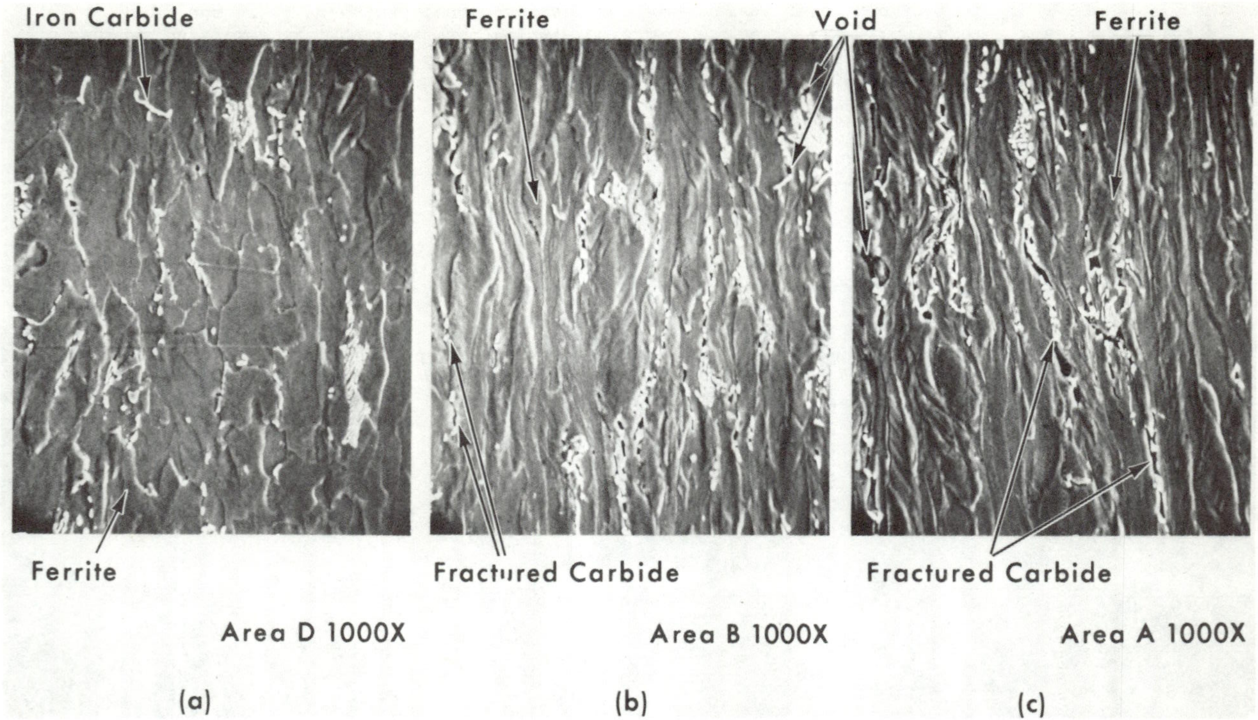

Fig. 12 - Scanning Electron Micrographs of deformed SAE 950X steel. (a) Elongation of the parent ferrite phase is evident. (b) The ferrite grains are grossly elongated and some iron carbides are fractured forming voids at the ferrite-carbide interface. (c) A larger number of voids are present. Note similarity of these micrographs with Fig. 11.

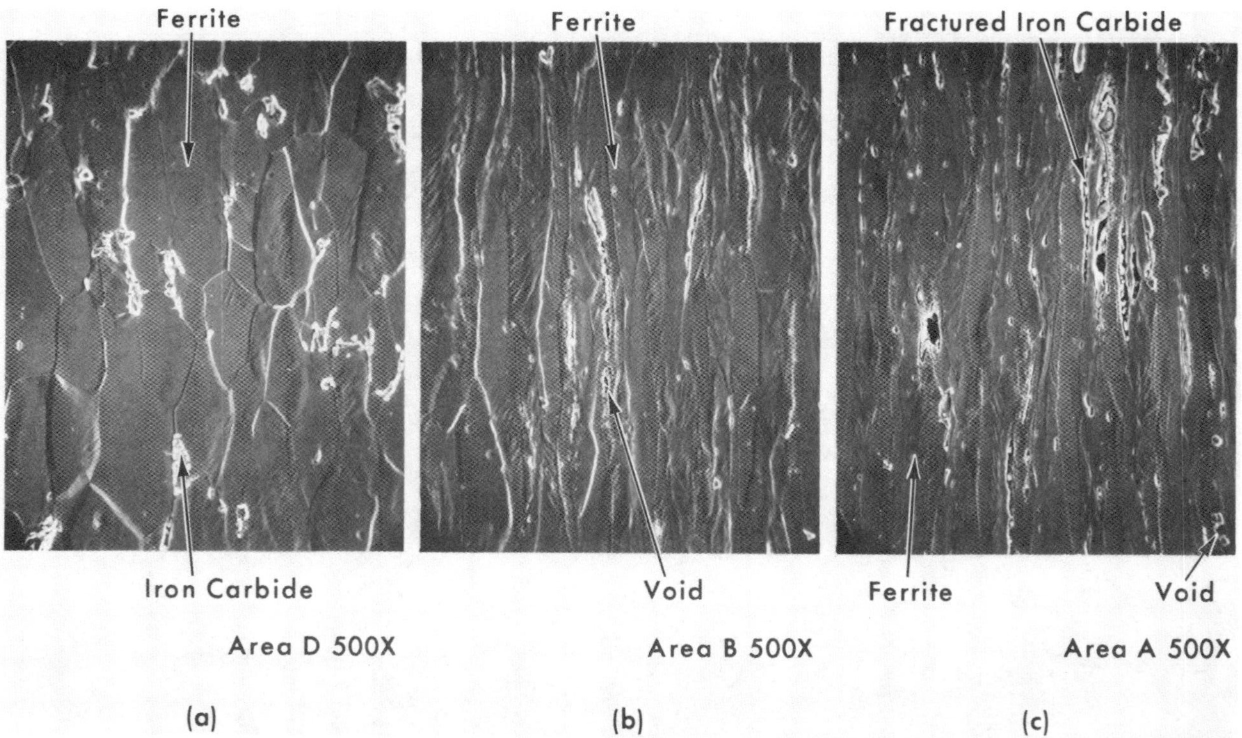

Fig. 13 - Scanning Electron Micrographs of Plain Carbon Steels. (a) Ferrite grains are elongated. (b) The elongation is more pronounced and some voids are present at grain boundaries. (c) A large number of voids and cracked iron carbides are present. Although the volume fraction of carbides is much less these micrographs are quite similar to those in Figs. 11 and 12.

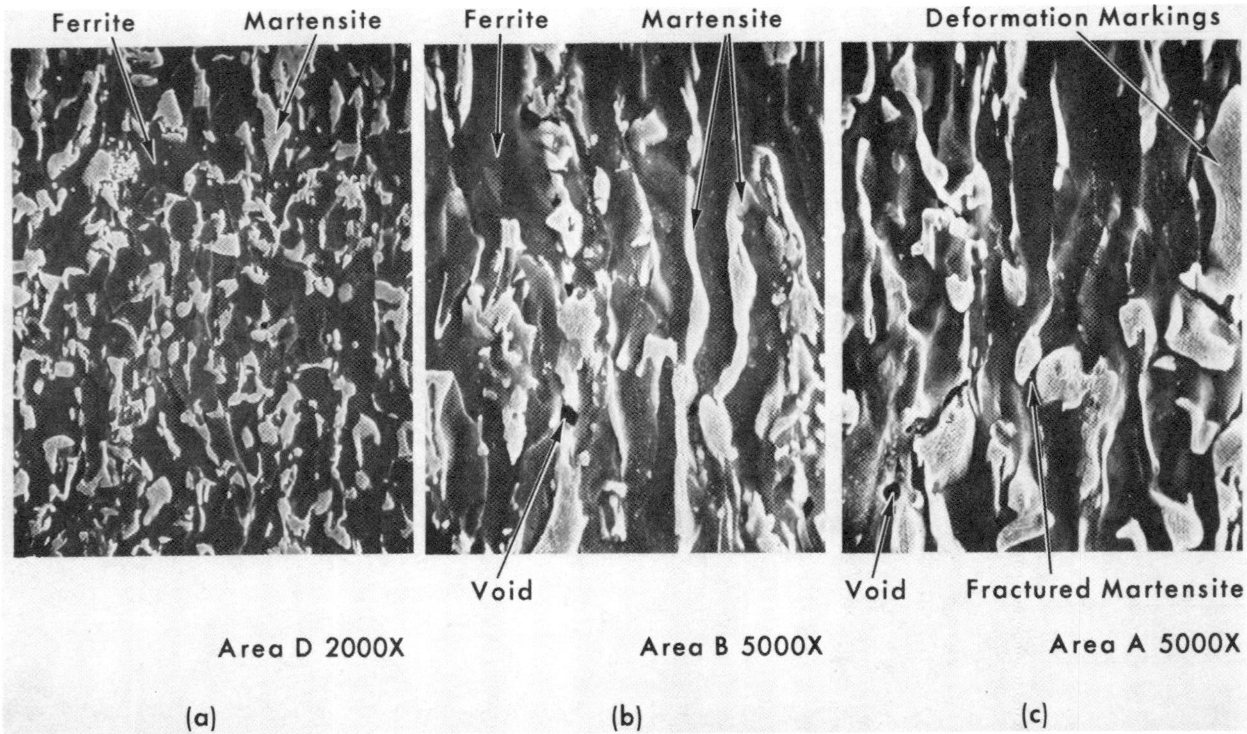

Fig. 14 - Scanning Electron Micrographs of deformed GM 980X. (a) Some elongation of the ferrite and martensite phases is evident. (b) Elongation of both phases is more pronounced and some voids are present at the ferrite-martensite interface. (c) Deformation markings are present in the elcngated martensite. A larger number of voids and fractured martensite islands are present than in (b).

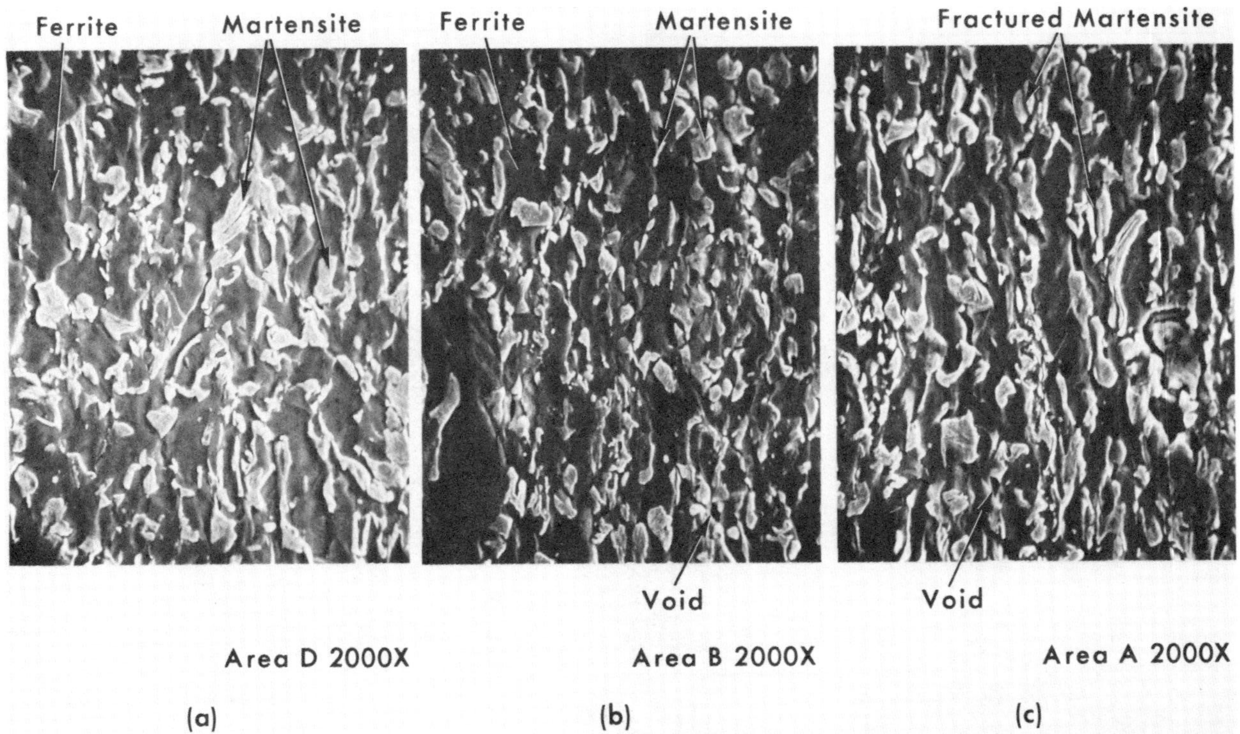

Fig. 15 - Scanning Electron Micrograph of heat treated SAE 980X steel with cooling rate D. (a) Some elongation of the ferrite and martensite phases is evident. (b) Elongation of martensite is not as pronounced as in Fig. 14b. Some voids are present at the ferrite-martensite interface. (c) A larger number of voids and fractured martensite islands are present than in (b).

DISCUSSION

Strain Hardening

All specimens tested exhibited a considerable variation in n_i for strains smaller than 0.05 (Figs. 8, 9). This variation is dependent upon the mechanisms by which initial plastic yielding occurs [7]. The trends for the as-received, ferrite-carbide and the heat treated ferrite-martensite specimens were different, however, suggesting that mechanisms by which initial plastic yielding occurs in these two groups of steels must be different. These mechanisms and the higher initial n_i values in the heat treated steels are most probably related to the nature of the parent matrices of the steels. Namely, as a result of the martensite transformation the ferrite matrix of the heat treated steels has a much larger density of mobile dislocations [1] than the corresponding as-received steel, and the distribution of VCN precipitates in the ferrite matrix before and after heat treatment is quite different.

For strains larger than 0.05, n_i remained practically constant in the three as-received steels (Fig. 8) at a value numerically equal to the full curve n value at these strains (Fig. 5). The σ-ε data for these steels can, therefore, be approximated by Eq. 1.

Log σ-log ε_p data for all the heat treated specimens (Fig. 7) had continuously changing slopes. As a result, n_i was not constant but, after an initial sharp increase, decreased continuously with increasing strain (Fig. 9). When the log σ-log ε_p data were approximated by straight lines, their resultant slopes or n values were different from their respective ε_u (Table III) obtained in the tension test. Therefore, the σ-ε data for the tested ferrite-martensite steels cannot be approximated by Eq. 1. Apparently, the strain hardening mechanism which becomes operative in the as-received steels after initial plastic yielding remains operative until the onset of necking, while in the heat treated specimens this is not so.

All the heat treated specimens had a ferrite-martensite or dual phase microstructure. However, for the chemistry and heating cycles evaluated, only specimens with the three slower cooling rates had $e_t >$ 27%. At the higher cooling rate, YS and UTS were similar to that of the three slower cooling rates, but e_t was lower and tended toward that for the SAE 980X steel (Fig. 6). The changes in microstructure which may be responsible for the lower e_t and possibly lesser formability are being investigated. Furthermore, specimens with cooling rates A and C produced identical mechanical properties except that their strain hardening behavior was different. This may influence formability and is also being evaluated.

The observed strain hardening behavior of the tested heat treated specimens also helps explain a previously perplexing question. Backofen showed [8] that the n value or true strain at necking - YS data for various steels fell within a narrow band (Fig. 16) which may be approximated by the relationship YS x n = 69 MPa (10,000 psi). The n values for the heat treated specimens (Fig. 7) do not fall within this band. This is now understandable because the heat treated specimen σ-ε data cannot be approximated by Eq. 1. The relationship is satisfied, however, when the value of n_i at the onset of necking is used instead of n (Fig. 16).

Microstructural Deformation

The elongation of ferrite grains was visible at all levels of deformation in all steels (Figs. 11 to 15) examined. Deformation of other constituents was less discernable in areas of low strain, and this had to be studied after gross deformation, such as that which occurs within a locally necked region. Plastic deformation in the as-received steels appeared to be limited to the ferrite matrix with minimal deformation of the iron carbide particles. Deformation occurred in the heat

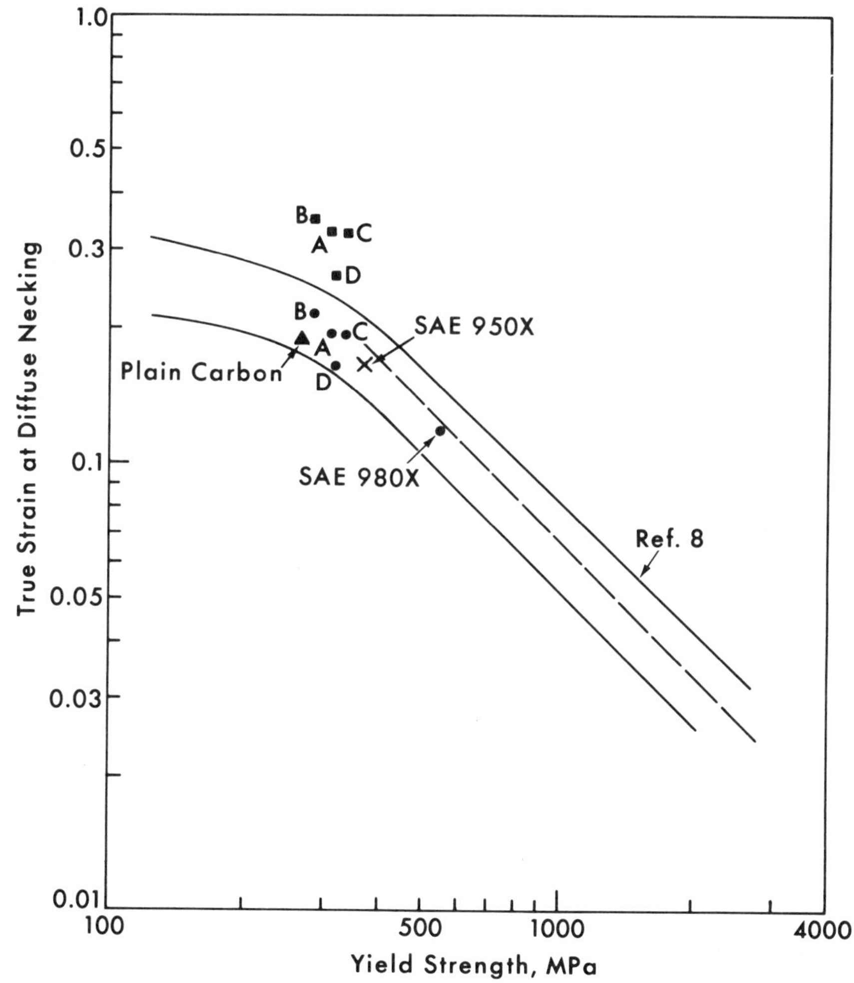

Fig. 16 - n_i values at the onset of necking (●) for all specimens tested fall within the band plotted from Ref. 8. The n value or slopes of the linear portions of the log σ-log ε_p plots (■) taken from Fig. 7 do not.

treated specimens which had $e_t > 27\%$ and met the GM 980X specification by plastic flow in both the ferrite matrix and the martensite islands (Fig. 14b). Evidently, deformation occurs first in the ferrite which is the continuous phase. While reaching its maximum capacity for being strained the ferrite matrix obviously transfers strain across the ferrite-martensite interface leading to deformation of the martensite islands. Since the martensite islands deform in these steels unlike the rigid iron carbide particles in the SAE 980X steel, localized necking does not occur in GM 980X until it is deformed more extensively than the SAE 980X steel. Therefore, strain is distributed more uniformly in the former steel and it has a greater resistance to necking than the latter steel and hence has far superior formability. Failure is initiated in the as-received steels by void formation in the highly strained ferrite around the iron carbide particles and by fracture of the iron carbides. The fracture is propagated by coalescence of these voids (Figs. 11c, 12c, 13c). In GM 980X failure is apparently delayed until both the ferrite and martensite phases are highly strained (Fig. 14b) and finally occurs by coalescence of voids formed at the ferrite-martensite interface and by fracture of the martensite islands.

The microstructure of the heat treated specimen with cooling rate D was similar to that of the three slower cooling rates. However, the volume fraction of martensite was somewhat higher and deformation of these martensite islands was not immediately obvious (Fig. 15). This martensite was probably less compliant and may be a cause of the observed lower e_t. Differences in martensite volume and characteristics at the various cooling rates are being evaluated.

The SEM observations appear to explain the strain hardening behavior of the steels. Only deformation of the ferrite phase was noticed in the as-received steels. The constant n_i for strains larger than 0.05 would, therefore, represent the strain hardening coefficient of the ferrite phase. Deformation of the ferrite and martensite phases were observed in the heat treated steels with the ferrite-martensite or dual phase microstructures. The measured n value would, therefore, represent the combination of the n values of both phases. However, n_i changed continuously with increasing strain suggesting that if both phases deform simultaneously their work hardening behavior must vary with increasing strain or different volume fractions of each phase probably become operative at different levels of strain, the latter being more likely.

The superior formability of GM 980X compared to SAE 980X steels [6] appears to depend, therefore, upon the nature of two microstructural constituents, a ferrite matrix with a very high strain hardening coefficient, and a deformable martensite phase. Obviously, the exact nature of these constituents must be important and variations in the nature of these constituents could influence steel formability. This is the subject of an ongoing investigation.

All dual phase steels are not equal. Steels have been reported [3] with UTS in the range 300 to 700 MPa (45,000 to 100,000 psi) with accompanying variations in e_t and formability. Also, different measuring techniques, specimen geometries [9] and gage lengths have been used to determine e_t. Therefore, in order to be valid, steel mechanical property comparisons should be made only when the data is obtained on identical specimens using similar techniques and when the steels have the same UTS.

CONCLUSIONS

1. Ferrite-martensite or dual phase steels produced by heat treating certain compositions of a ferrite-carbide SAE 980X steel have significantly higher strain hardening coefficients than the SAE 980X steel over the entire strain range examined.

2. True stress-true strain data for the tested ferrite-martensite or dual phase steels can not be approximated by the relationship $\sigma = K\varepsilon_p^n$ while data for examined as-received steels with ferrite carbide microstructures can.

3. Previously the product of yield strength and n has been considered to be approximately the same for all steels. However, this relationship is satisfied by the tested dual phase steels only when the incremental strain hardening coefficient, n_i, at maximum load is substituted for n.

4. The ductility of the dual phase steels is affected by the heat treatment cooling rate. For the composition investigated, and for tensile strengths within the range 650 to 700 MPa, cooling rates between 5.2° and 11.8°C/sec resulted in total elongations larger than the 27% minimum specified for GM 980X.

5. Plastic deformation takes place in the as-received, SAE 980X, 950X and plain carbon steel by straining of the parent ferrite phase with practically no deformation of the grain boundary iron carbides. Deformation in GM 980X is controlled by strain hardening of both the ferrite and martensite phases. The observed constant n_i in the asreceived steels is, therefore, the strain hardening coefficient of the ferrite matrix. In GM 980X, the observed continuously varying strain hardening coefficient is apparently a combination of the coefficients of the two deforming phases.

6. GM 980X has better formability and a greater resistance to necking than the SAE 980X steel because, in part, the GM 980X ferrite matrix has a high strain hardening coefficient, and because its martensite phase is deformable. When the SAE 980X steel ferrite is highly strained, voids form preferentially at ferrite-iron carbide interfaces and fracture is initiated. When the GM 980X ferrite is highly strained, strain is apparently transferred to the martensite phase and it also deforms. Therefore, voids leading to failure do not form until after more extensive deformation has occurred and the martensite phase is also highly strained.

ACKNOWLEDGMENTS

The expert assistance of Messrs. Harold G. Lemmer, Carl Ligotti and Erwin R. Cprek in all experimental aspects of this work is sincerely appreciated.

REFERENCES

1. M. S. Rashid: SAE Preprint 760206, Feb. 1976.

2. M. S. Rashid: SAE Preprint 770211, Mar. 1977.

3. S. Hayami and T. Furukawa: Micro Alloying 75, Proceedings of the Conference, pp. 311-320, Micro Alloying 72, 270 Park Avenue, New York, 1977.

4. S. P. Keeler: series of six articles, Machinery, 1968.

5. G. E. Dieter, Jr.: Mechanical Metallurgy, pp. 56, McGraw Hill Book Co., New York, 1961.

6. M. S. Rashid and E. R. Cprek: ASTM Symposium on Formability, Proceedings, to be published, 1977.

7. L. A. Erasmus: Met. and Met. Forming, vol. 42, pp. 94-99, 1975.

8. W. A. Backofen: Massachusetts Institute of Technology Industrial Liason Symposium, Chicago, 1974.

9. T. Matsuoka and K. Yamamori: Met. Trans., vol. 6A, pp. 1613-22, 1973.

APPENDIX

Calculation of n_i

A number of marks are made on the true stress-true strain (σ-ε) curve, usually at equal intervals of strain. The marks are indexed 0, 1, 2, 3,.. j .. m with 0 being the proportional limit for a smooth σ-ε curve and m the point of fracture. When yield point elongation (ype) is present, the end of the ype plateau is taken as 0 instead of the proportional limit. The σ-ε coordinates of each mark are measured. The incremental strain hardening exponent, n_i, at each mark is then calculated as follows:

$$n_i = \frac{\log \sigma_j - \log \sigma_{j-1}}{\log \varepsilon_{p_{(j)}} - \log \varepsilon_{p_{(j-1)}}} \qquad j \rightarrow 1 \text{ to } m$$

ON THE DUCTILITY OF DUAL PHASE STEELS

R. G. Davies
Research Staff
Ford Motor Company Dearborn, Michigan

The factors that influence both the strength and ductility of dual-phase (essentially ferrite plus martensite) steels are reviewed. It is found that the strength is dependent primarily upon the volume fraction of martensite in the structure; ferrite grain size and solid solution additions also contribute to the strength. The ductility of dual-phase steels is largely a consequence of the high strength (fine grained, solid solution strengthened), highly ductile (low interstitial content) ferrite matrix. The strength and ductility are shown to be in agreement with Mileiko's theory of composites of two ductile phases. From a study of the influence of Si and P on the mechanical properties of iron it is shown that the theory is able to predict the properties of Si and P containing dual-phase steels. The addition of Si is found to be beneficial as, in addition to decreasing the interstitial content of the ferrite, it increases the strength more than it decreases the ductility. The necessity of "clean" ferrite for optimum properties is confirmed by comparing ferrite-plus pearlite dual-phase steels with normalized steels

I. Introduction

Dual-phase steels, so called because they consist essentially of a mixture of 15-20% martensite in a ferrite matrix, are of great interest because they offer a unique combination of strength and ductility (1-3). For example, the dual-phase steel VAN-QN* has a tensile strength of about 650 MPa (94 ksi) and a total elongation of 28% (4), whereas a conventional 980 HSLA steel of similar tensile strength has a total elongation of only about 20%. The increased elongation of the dual-phase steels translates into better formability at a given tensile strength (2), and thus these steels could be important where the maximum weight reduction is being sought in stamped parts.

In this paper we will discuss how factors such as ferrite grain size, solid solution strengthening and ductility, and martensite strength and volume fraction influence both the strength and ductility of dual-phase steels. It will be shown that the strength and ductility of the ferrite is the major factor governing the mechanical properties of dual-phase steels; the ferrite should be fine grained and have a low interstitial content. It is found that a theory of a composite of two ductile phases (5,6) offers a good description of the strength and ductility of these dual-phase steels.

II. Strength of Dual-Phase Steels

A study has been made of the effect of volume fraction of martensite upon the strength of Fe-1.5 Mn-C dual phase steels (7). The percent martensite in the structure was varied by quenching a series of different carbon-content alloys from a constant temperature; thus the composition of the ferrite and martensite in the dual-phase structures remains constant. In addition the alloys were quenched from other temperatures in the intercritical region so as to vary the composition of the martensite. It was found, as shown in Figure 1, that the strength of these dual-phase steels is dependent only upon the volume fraction of the martensite present, and independent of the quench temperature, i. e., carbon content of the martensite. This is an unexpected result since, if a law of mixtures applies, then the flow stress of the alloys quenched from 740°C (carbon content in martensite $\sim$ 0.57%) should be measurably stronger than alloys quenched from 800°C (carbon content $\sim$ 0.38%); the strength of martensite is very sensitive to such changes in carbon content (8,9).

Extrapolating the 0.2% flow stress to 100% martensite gives a strength indicative of a 0.4% carbon martensite tested in tension (8); this strength is much lower than expected, especially for the alloys quenched from 740°C. A similar result was found by Tamura et al. (10) who quenched a series of Fe-C alloys from 785°C; they found that the strength of the dual-phase structures varied approximately linearly with percent martensite, and that the strength for the 100% martensite structure obtained by extrapolation was considerably below the measured strength of a 100% martensite alloy.

Possible reasons for the strength of dual-phase steels apparently being lower than indicated by a law of mixtures are: 1. auto-tempering in the higher carbon martensites, 2. a strain component from the transformation under stress of the retained austenite in the high carbon martensites, 3. a larger Bauschinger effect due to the lower transformation temperature and larger transformation strains in the higher carbon martensites, and 4. martensite has a lower strength when it is formed in a ferrite matrix. The transformation substructure, and therefore strength, of a martensite island formed in a soft ferrite matrix may be quite different from the substructure formed when a steel is transforming from austenite to 100% martensite. Austenite is generally stronger at high temperatures than ferrite and will therefore impose more severe constraints on the forming martensite regions. A careful transmission electron microscope study should be able to clarify this last point.

*A proprietary product of Jones and Laughlin Steel Corporation.

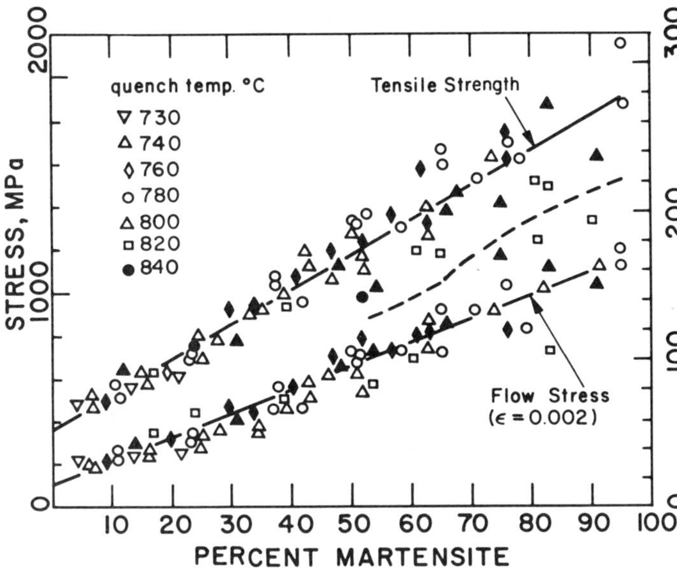

Fig. 1 The 0.2% flow stress and the tensile strength as a function of percent martensite for Fe-Mn-C alloys (7).

As the structure of dual-phase steels consists of 80-85% ferrite when the strength is about 690 MPa (100 ksi), it is not surprising that processes that increase the ferrite strength also increase the strength of dual-phase structures. However, as will be shown in the section on ductility, there must be a balance between the increase in strength and any accompanying loss in ductility of the ferrite. It has been found that ferrite grain refinement (4,7) and certain solid solution strengthening elements (11) can lead to useful increases in the strength of dual-phase steels. Thus the factors that control the strength of dual-phase steels appear to be well understood.

III. Ductility

III.1 Measures of Ductility

It has been known for many years that the total elongation of sheet specimens is dependent upon both the type of material and the thickness of the sheet, while the uniform elongation is relatively independent of sheet thickness (12,13). To evaluate this thickness effect in a dual-phase steel, a thick sheet (4 mm) of the steel was reduced in thickness by grinding off equal amounts from each side to produce a series of specimens whose thicknesses varied from 1.25 mm (0.05 in) to 3.75 mm (0.150 in); these, and all the samples whose data are presented here, had a 50 mm (2.00 in) gage length and a width of 12.5 mm (0.500 in). The elongations, both uniform and total, and post-uniform (total minus uniform), are shown as a function of specimen thickness in Figure 2; it can be seen that the total elongation decreases with decreasing specimen thickness. Since the uniform elongation is independent of specimen thickness over the range of thicknesses studied, it was decided to use only the uniform elongation as a measure of the ductility.

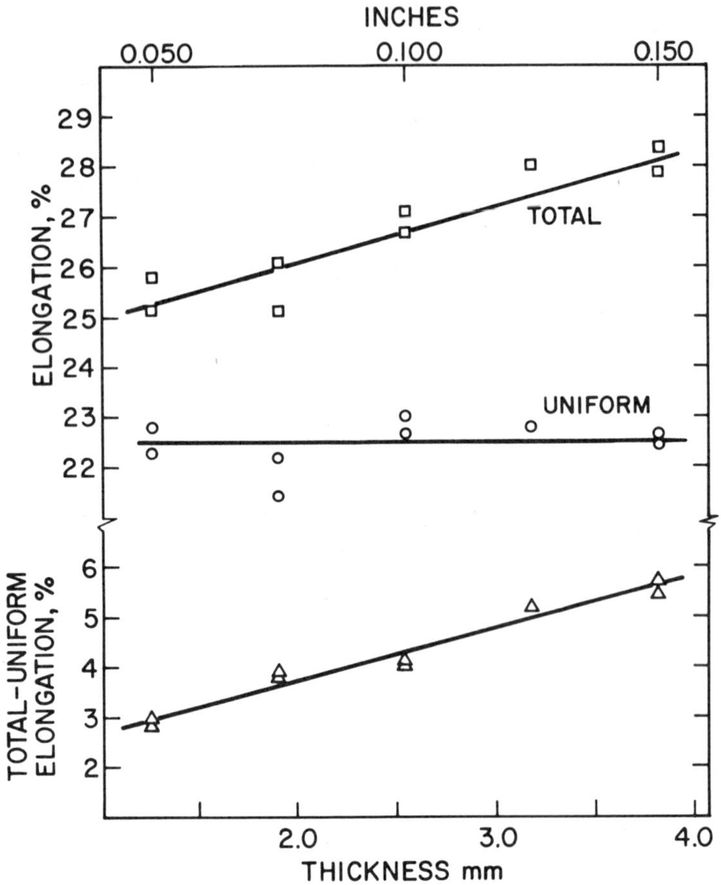

Fig. 2 The total and uniform elongations and post-uniform (total minus uniform) elongation as a function of specimen thickness for a dual-phase steel.

Another measure of ductility which is directly related to uniform elongation is n, the strain hardening exponent; this assumes that the material obeys the power law

$$\sigma = k\varepsilon^n \tag{1}$$

where σ is the true stress, ε the true strain and k is a constant. If Equation (1) is obeyed then n is equal to the true uniform strain. Since n decreases with increasing strain in dual-phase steels (14), the appropriate n value to compare to true uniform strain is that obtained at high strain values; for strains above 10%, n is approximately independent of strain. The good correlation between the n value determined at strains of greater than half the uniform strain and true uniform strain is shown in Figure 3 for several dual-phase steels (15). Thus it is possible and sometimes convenient to use n and true uniform strain interchangeably. It is often easier to quantify n than the uniform strain, because the determination of the uniform strain relies upon the detection of the maximum load and this is dependent upon the sensitivity of the load measuring system of the tensile testing equipment.

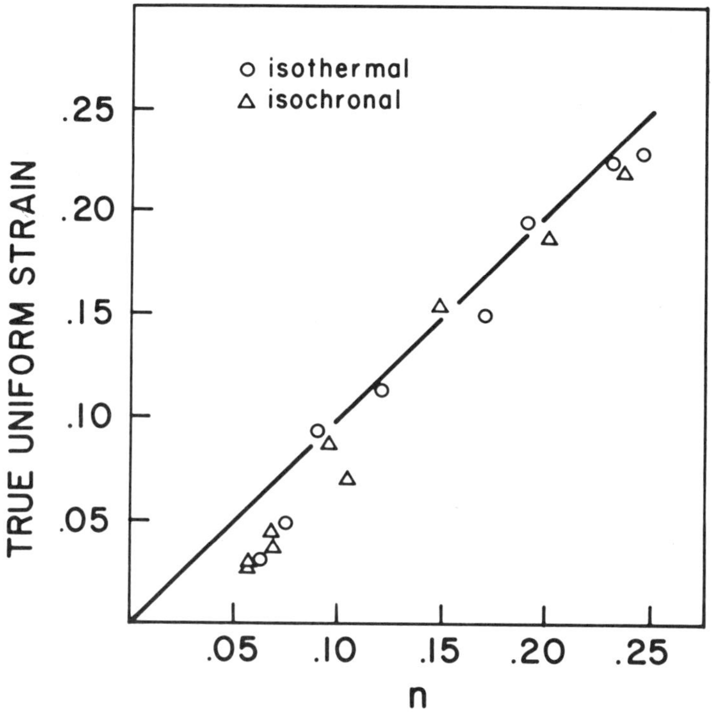

Fig. 3 This graph shows the good correlation between n and true uniform strain; the solid line is for perfect correlation (15).

III.2 Dual-Phase Steels

The uniform elongation as a function of tensile strength for the Fe-Mn-C dual-phase steels (7) and for conventional HSLA steels is shown in Figure 4, where it can be seen that at all strength levels the dual-phase steels exhibit superior elongations. Figure 5, which contains data for a vanadium (VAN-80) and a molybdenum containing dual-phase steel, shows that the enhanced ductility is not confined to a single alloy system. In addition it can be seen that at a given strength level, a finer grain size can further increase the elongation and that alloys containing nonmartensite products (bainites and fine pearlites) can also have better elongations than conventional HSLA steels. Thus it is clear that the superior ductility of dual-phase steels arises from the ferrite plus martensite structure produced by intercritical annealing and not from the presence of any specific alloying element.

III.3 Theory

It has been shown previously (3,7,15) that dual-phase steels obey quite well the theory of composites of two ductile phases as developed by Mileiko (5) and Garmong and Thompson (6). Many people do not think of high carbon martensites as being ductile;

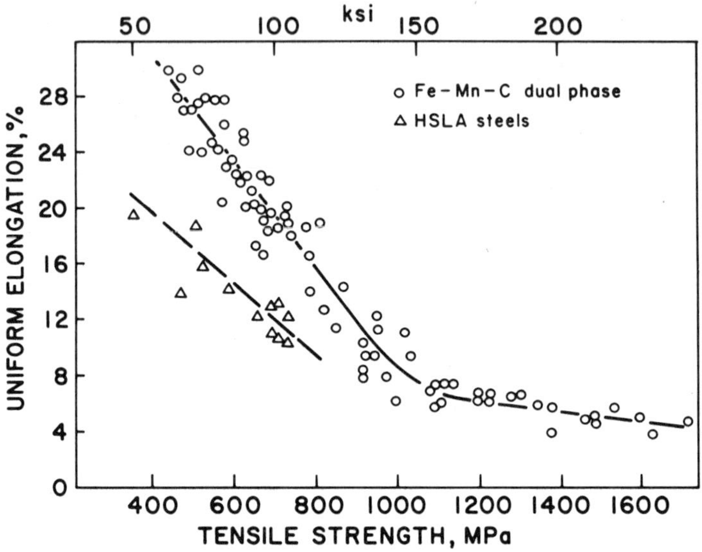

Fig. 4 Uniform elongation as a function of tensile strength for standard HSLA steels and Fe-Mn-C dual-phase steels (7).

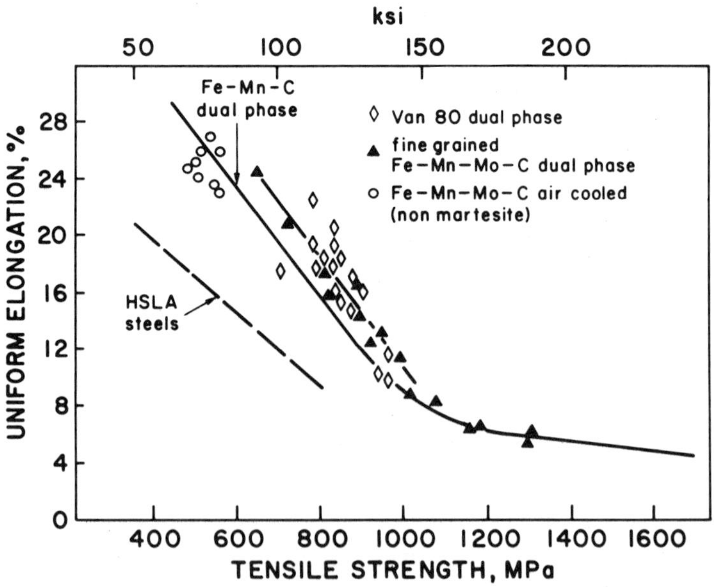

Fig. 5 Uniform elongation as a function of tensile strength (solid line for Fe-Mn-C is from Fig. 4) showing the influence of grain size and various alloying additions on the properties (7).

however as shown by Rashid (14), the martensite islands in a ferrite matrix undergo large deformations without fracture or interface separation. This theory involves three basic assumptions: (1) the two component phases and the composite obey the power law of Equation (1) (available data (2,4,15) indicates that dual-phase steels and their components are in reasonable agreement with this equation), (2) perfect interface matching between components (Koo and Thomas (16) have shown that the ferrite and martensite in dual-phase steels have perfect interfacial matching), and (3) the second phase to be in the form of aligned fibers (the martensite in dual-phase steels is not aligned but this basically geometric consideration may be unimportant for components with perfect interfacial matching so that all regions continue to carry load at all strain levels).

The theory predicts that, to a first approximation, the tensile strength of the composite is proportional to the amount of second phase (law of mixtures), and that the ductility will be less than that indicated by a law of mixtures. A study of an Fe-3Ni-3Mo alloy (15) where the properties of the ferrite, martensite and composite could be measured, showed that the theory provides a good description of the composite properties. Figure 6 shows good agreement between the calculated and measured values of n (or true uniform strain) as a function of percent martensite. Similar agreement was obtained for the Fe-Mn-C dual-phase steels (7).

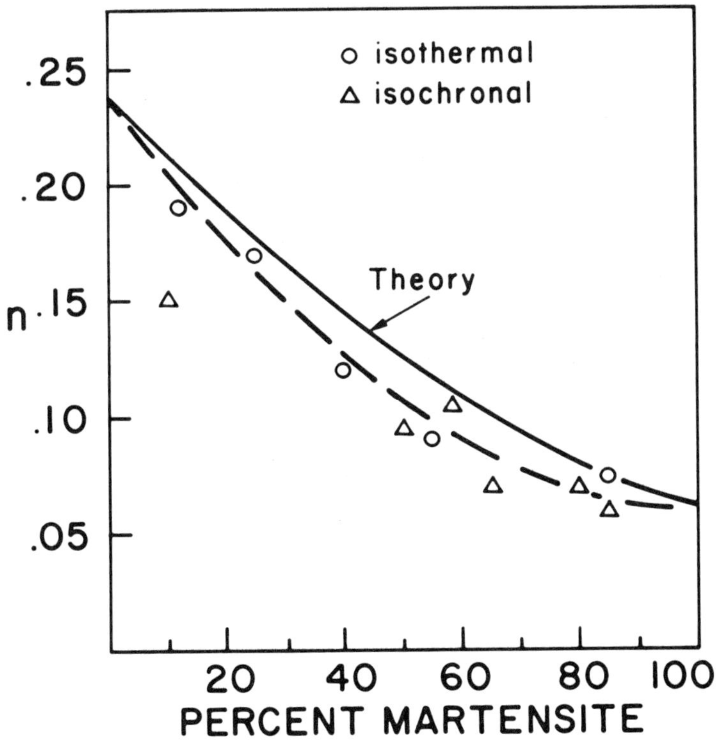

Fig. 6 Strain hardening exponent n as a function of percent martensite compared to the theoretical prediction (15).

Since the theory of composites appears to describe the results for dual-phase steels quite well, it is possible to calculate the changes in ductility with tensile strength of the composite for various combinations of martensite and ferrite properties. It was found (3) that at a composite strength of 690 MPa (100 ksi) there would be only a small increase in the ductility if the martensite tensile strength could be increased from 1380 MPa (200 ksi) to 2620 MPa (380 ksi). On the other hand increasing the tensile strength of the ferrite from 276 to 552 MPa (40 to 80 ksi), all other factors held constant, would as shown in Figure 7, lead to a large increase in the ductility of the dual-phase steel at a strength of 690 MPa (100 ksi).

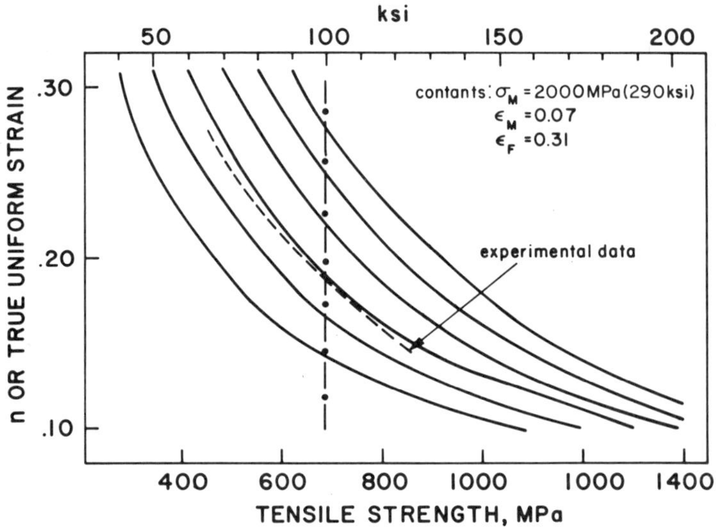

Fig. 7 The predicted n or true uniform strain as a function of tensile strength for ferrites with strengths varying from 275 to 620 MPa (40 to 90 ksi) (7).

The above analysis indicates that to achieve the greatest ductility at a given tensile strength, the dual-phase steel should be composed of strong martensite in a matrix of the highest strength, highest ductility ferrite possible. The requirement of high strength and high ductility in the ferrite may be impossible to attain; ductility usually decreases as strength increases. Morrison (17) found that only for pure Fe was the ductility independent of grain size (i.e. strength); any carbon or carbides in the alloys led to a decrease in ductility with decrease in grain size. Thus it has been proposed (3) that for optimum properties of dual-phase steels, very fine grained ($< 5\mu m$ diameter), clean (low interstitial content) ferrite is required.

Another means of increasing the ferrite strength is by substitutional solid solution strengthening; two of the best ferrite strengtheners are P and Si (18). However, it was not possible to assess the influence of P and Si on the properties of dual-phase steels since there is no data on the influence of P and Si on the strength and ductility of iron at very fine grain sizes. Thus a study has been made of the influence of P and Si on the properties of both iron and dual-phase steels (11); the results are summarized in the following section.

IV. Solid Solution Strengthening

IV.1 Ferrites

A series of Fe-Si and Fe-P alloys with carbon contents < 0.01% were prepared by vacuum melting. The ingots were hot-rolled to a thickness of 15 mm (0.60 in) and then cold-rolled to 1.5 mm (0.060 in) thick sheet. A range of grain sizes was produced by recrystallizing at various temperatures between 650 and 950°C; standard tensile samples with a 50 mm (2.0 in) gage length and 12.5 mm (0.50 in) wide were prepared from this material and tested at a cross-head rate of 0.04 mm/s.

The true tensile strength and n (ductility) as a function of grain size are shown in Figures 8 and 9, respectively, for both the Si and P containing alloys. The results show: (1) confirmation of Morrison's observation that for pure iron the ductility is approximately independent of grain size, (2) for all the alloys the strength increases and ductility decreases as the grain size is decreased, and (3) significant solid solution strengthening with the 0.2 wt. % P alloy being as strong as the alloy containing 2 wt. % Si.

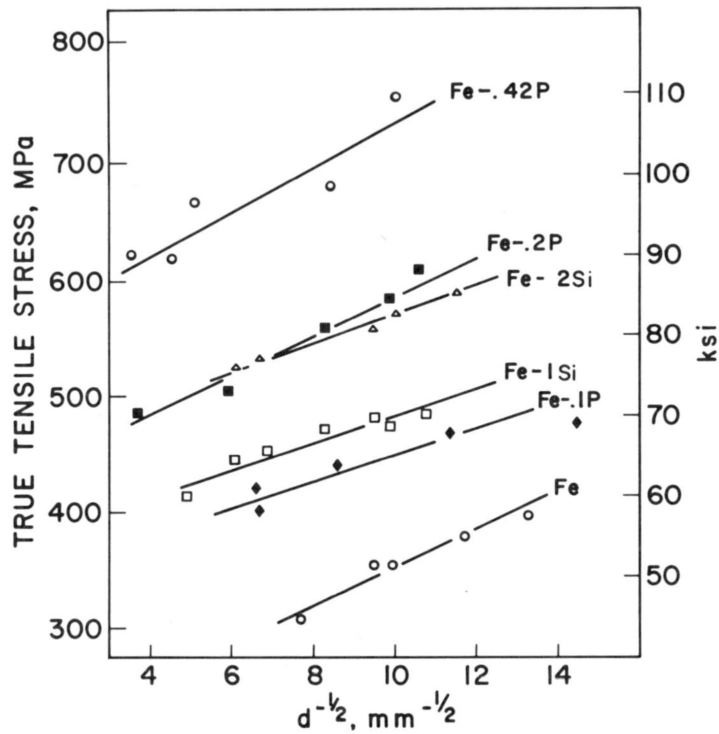

Fig. 8 Tensile strength as a function of grain size for several Fe-Si and Fe-P alloys (11).

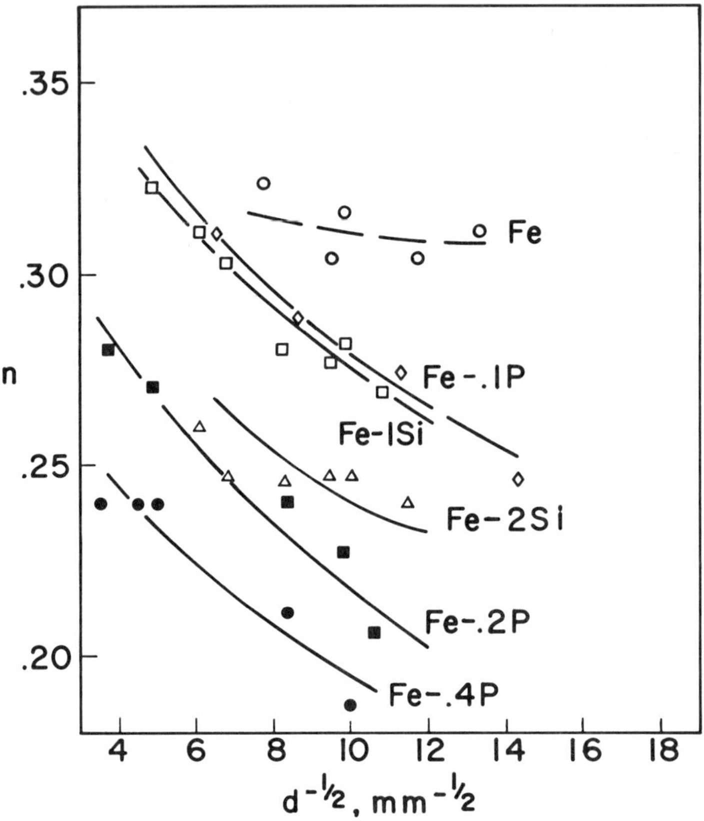

Fig. 9 Strain hardening exponent n as a function of grain size for several Fe-Si and Fe-P alloys (11).

IV.2 Dual-Phase Steels

With the above strength and ductility data it is now possible to calculate the expected properties of dual-phase steels containing Si and P. The ferrite properties were taken at a grain size of $d^{-1/2}$ $mm^{-1/2}$ of 12 (ASTM 11-12) since this is the grain size often found in dual-phase steels (2-4), and it is assumed that the martensite properties are unaffected by Si or P and the same as in Figure 7. The computed properties of these dual-phase steels, Figure 10, shows that steels containing 0.1 or 0.2 wt. % P should have similar ductilities and that those steels are inferior to both the Si containing steels; the steel containing 2 wt. % Si should have the best properties.

To test these predictions, a series of Si and P containing dual-phase steels were produced by vacuum casting (11); these steels all contained 1.5 wt % Mn for hardenability and 0.05-0.10% C to allow for the production of structures with 10-20% martensite. These alloys in the form of 1.5 mm (0.06 in) thick sheets were intercritically annealed for 10 minutes at temperatures between 740 and 840°C and either brine quenched or air cooled. The results for the Si containing alloys are shown in Figure 11

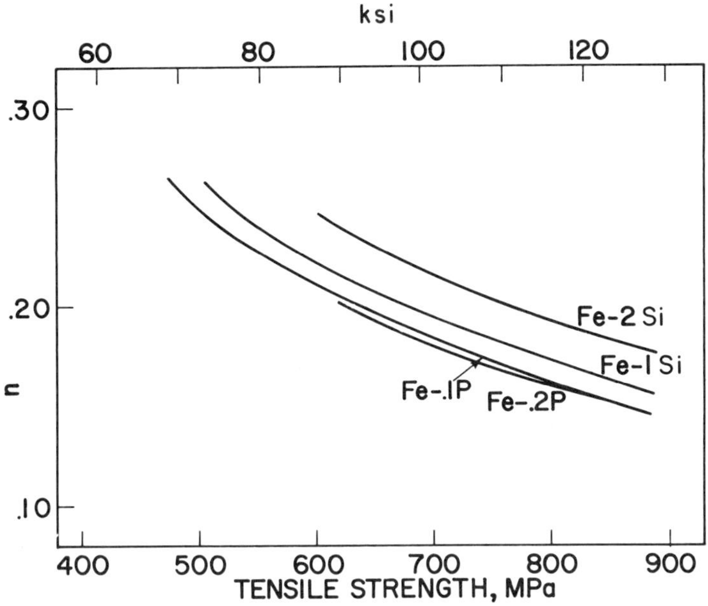

Fig. 10 Calculated n or true uniform strain as a function of tensile strength for Si and P containing dual-phase steels (11).

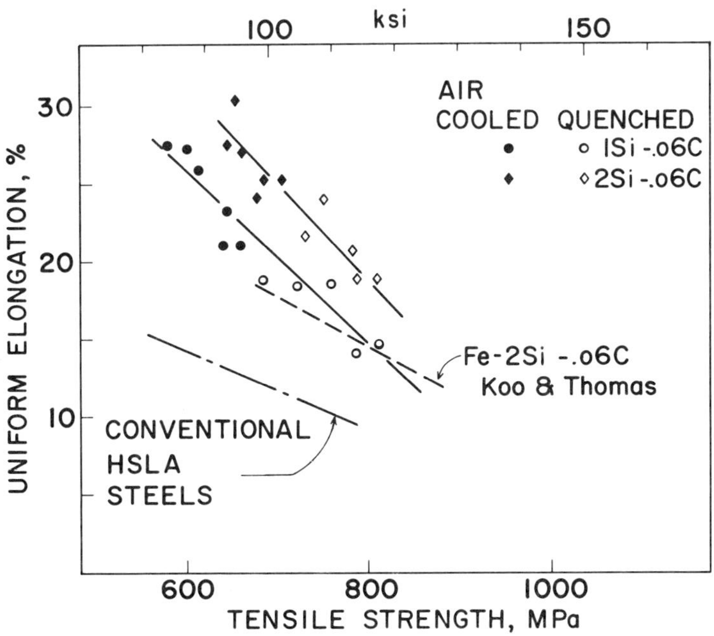

Fig. 11 Uniform elongation as a function of tensile strength for Si containing dual-phase steels (11).

where it can be seen that the 2% Si alloy is, as predicted, superior to the 1% Si alloy, and that there is a large difference in uniform elongation between the two alloys containing 2% Si (present work (11) and Koo and Thomas (20)). Some of this difference could be a result of the specimen geometry; Koo and Thomas (20) pulled 6 mm (0.25 in) diameter specimens which will result in a triaxial stress state while the present sheet samples will be in a mainly biaxial stress state.

The results for all the current Si and P containing dual-phase steels are collected together in Figure 12 (the data points have been left out for the sake of clarity); except for the reversal in order for the 0.1 and 0.2 P alloys, the results are as predicted in Figure 10. This agreement between prediction and experiment gives added confidence to the correctness of treating dual-phase steels as a composite of two ductile phases. It is worth noting at this point that Si in a dual-phase steel has other beneficial effects besides increasing the strength/ductility combination; for example, it will increase the hardenability and make it easier to obtain martensite, but more importantly Si reduces the carbon content of ferrite which will lead to a "cleaner" more ductile ferrite.

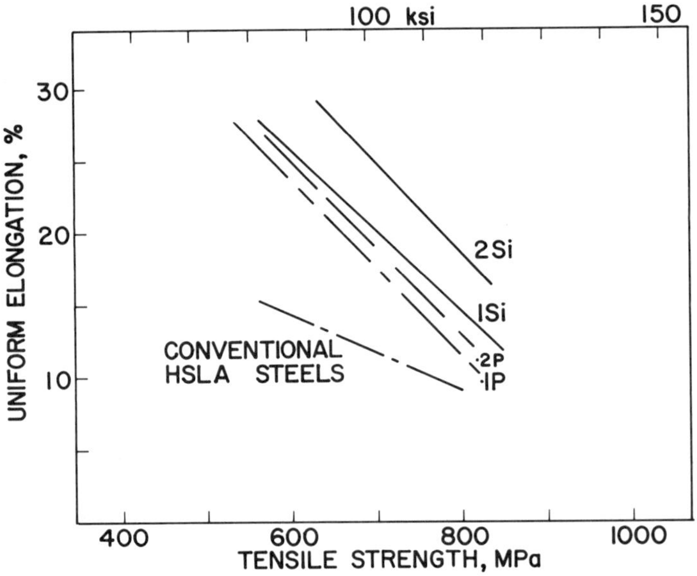

Fig. 12 Uniform elongation as a function of tensile strength for both Si and P containing dual-phase steels; data points have been omitted for clarity (11).

IV.3 Dual-Phase Pearlite Steels

Although the discussion so far has revolved around ferrite-martensite dual-phase steels, there is no reason why mixtures of other types of phases should not behave in a similar manner. It was shown in an earlier publication (3) that a dual-phase ferrite-pearlite steel had greater elongation than a conventional HSLA steel of comparable tensile strength. To confirm this observation some of the Fe-Mn-C alloys used to generate the data of Figure 4 were heat-treated as follows: (1) to produce dual-phase pearlite - intercritically anneal at 740°C for 10 minutes and quench into a salt bath at 575°C and hold for 15 minutes to transform the austenite to fine pearlite, and (2) to give a normalized structure - fully austenitize at 900°C for 5 minutes and air-cool.

The uniform elongation as a function of tensile strength for the above two heat-treatments are compared in Figure 13 to data for conventional HSLA and ferrite-martensite dual-phase steels. It can be seen that the dual-phase ferrite-pearlite steels are not quite as ductile at a given tensile strength as the ferrite-martensite steels; this is a consequence of the pearlite having a lower strength than the martensite and the result is consistent with the theory.

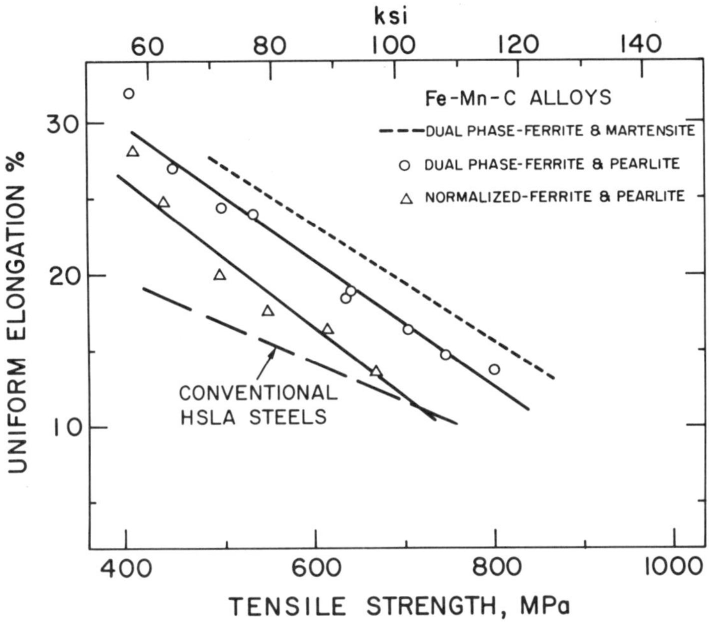

Fig. 13 Uniform elongation as a function of tensile strength for various Fe-Mn-C structures and conventional HSLA steels.

It is important to note that the ductility is greater for the intercritically annealed than for the normalized steels, due no doubt to the differences in ferrite properties. These results again emphasize the need for a precipitate-free, low interstitial content ferrite to obtain the optimum properties in a dual-phase steel. The strength as a function of percent pearlite for the dual-phase ferrite-pearlite steels is shown in Figure 14; the tensile strength of 100% pearlite of 1034 MPa (150 ksi) is approximately half that for 100% martensite (see Figure 1)

V. General Comments

It is now quite clear that the superior properties of dual-phase steels arise from essentially structural features; the enhanced ductility at a given tensile strength is a consequence of the dispersion of a strong martensite in a high strength, highly ductile ferrite matrix. There are many combinations of alloying elements such as Mn, Si, Cr, Mo and V that can be added to low carbon ($\sim$0.1%) iron to obtain the desired structure and thus it is very unlikely that all steel producers will ever arrive at a single dual-phase alloy; every steel producer has different melting, rolling, annealing and cooling facilities and they will choose the alloying elements best suited to their production capabilities. In addition it has been shown (19) that it is possible to produce a whole family of dual-phase steels with various combinations of strength and ductility.

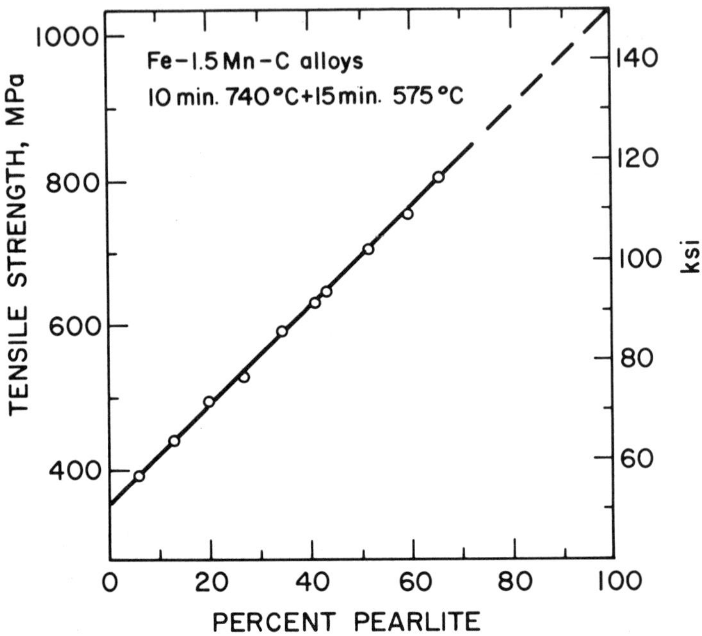

Fig. 14 Tensile strength as a function of percent pearlite in Fe-Mn-C alloys.

To avoid the confusion for the user of each producer having his own trade name/description for dual-phase steels, I believe we should as quickly as possible arrive at an agreed generic name/description for these steels. Even though it has been shown in this conference that these steels may consist of three or more phases, these steels are widely referred to, and people recognize them, as dual-phase steels; therefore why not call them DP steels. As the tensile strength and total elongation appear to be the important parameters to the materials engineer, I propose that to convey the maximum information, these steels could be described generically as DP xxx/yy, where xxx is the tensile strength in MPa or ksi and yy is the total elongation; for example, the J. and L. VAN-QN steel would be DP 90/27 (minimums of 90 ksi tensile strength and 27% total elongation).

References

1. S. Hayami and T. Kurukawa, Proc. Microalloying Conf., Washington, D.C., 1975, pp. 78-87.

2. M. S. Rashid, SAE Preprint 760206, February 1976.

3. R. G. Davies, Met. Trans., 1978, Vol. 7A, pp. 41-52.

4. J. H. Bucher and E. G. Hamburg, SAE Preprint 770164, February 1977.

5. S. T. Mileiko, J. Mat. Sci., 1969, Vol. 4, pp. 974-977.

6. G. Garmong and R. B. Thompson, Met. Trans., 1973, Vol. 4, pp. 863-873.

7. R. G. Davies, Met. Trans.. 1978, Vol. 7A, May.

8. W. C. Leslie, R. J. Sober, Trans ASM, 1967, Vol. 60, pp. 459-484.

9. C. L. Magee and R. G. Davies, Acta Met., 1972, Vol. 20, pp. 1031-1043.

10. I. Tamura, Y. Tomota, A. Akai, Y. Yamaoka, M. Ozawa, and S. Kanatani, Trans. I S I Japan, 1973, Vol. 13, pp. 283-292.

11. R. G. Davies, unpublished research.

12. R. L. Templin, Proc. Am. Soc. Testing Materials, 1926, Vol. 26, Part 2, pp. 378-398.

13. E. B Kula and N. H. Fahey, Materials Research & Standards, 1961, Vol. 1, pp. 631-636.

14. M. S. Rashid, this conference.

15. R. G. Davies, Met. Trans., 1978, Vol. 7A, March.

16. J. Y. Koo and G. Thomas, this conference.

17. W. B. Morrison, Trans. ASM, 1966, Vol. 59, pp. 824-846.

18. G. F. Bolling and R. H. Richman, Canadian J. of Phys., 1967, Vol. 45, pp. 541-557.

19. S. Hayami, T. Kurukawa, H. Gondok and H. Takechi, this conference.

20. J. Y. Koo and G. Thomas, Met. Trans., 1977, Vol. 8A, pp. 525-528

DESIGN OF DUPLEX LOW CARBON STEELS

FOR IMPROVED STRENGTH: WEIGHT APPLICATIONS

J. Y. Koo and G. Thomas

Materials and Molecular Research Division
Lawrence Berkeley Laboratory
and
Department of Materials Science and Mineral Engineering
University of California
Berkeley, California 94720

Design principles for improved mechanical properties of Duplex Ferrite-Martensite (DFM) steels have been evaluated in order to obtain desirable microstructural characteristics which in turn result in desirable mechanical properties. A DFM alloy, of composition Fe/2% Si/0.1% C, has been developed according to the design criteria. This 2% Si DFM steel showed high strength and good formability, and its tensile properties are superior to a series of Cr and Mn containing DFM steels, and to some selective commercial HSLA steels including Van 80. Over the range of 15 $\sim$ 80% martensite the "composite" rule for two phase mixtures appears to hold as a fairly good approximation of the tensile behavior of the duplex systems investigated.

The effect of the property variations of the individual constituents, and duplex structure-property relations are also presented.

Introduction

During the past several years, the duplex ferrite-martensite (DFM) steels have received increasing emphasis, especially in the transportation industries, due to the characteristic microstructural features which combine high strength with good formability. The strengthening principle of the DFM structure involves the incorporation of inherently strong martensite as a load carrying constituent in a soft ferrite matrix. The latter supplies the system with the essential ductility.

Apart from these interesting mechanical properties, however, there is a general lack of fundamental understanding of the characteristic behavior of DFM steels. For instance, papers reporting the results of mechanical property tests generally made little direct correlations with duplex microstructures, except for some occasional optical metallography (1-3). The mechanical behavior of the two phase materials as well as the intricate interactions of parameters such as the size, shape, distribution, and volume fraction of martensite particles, must all be characterized and controlled in designing improved DFM alloys such that they favorably contribute in concert to the overall mechanical properties.

In the present paper, the principles of our current alloy design program, which has been adopted to achieve optimum balances of strength and ductility, are briefly described.

Principles of DFM Alloy Design Program

The main principles of the alloy development can be summarized as follows:

1. Obtain the (α+ martensite) duplex structures by phase transformation alone.

2. Control the % C in martensite to $\sim$0.3% to avoid twinned martensite.

3. Alloys must have a large slope in the A_3 line.

4. Optimize the properties of the constituent phases.

DFM structures can be produced in many different ways (4), as has been demonstrated over the last 40 years (5-9). We restrict the present approach to employ only simple heat treatment by annealing in the two-phase ($\alpha+\gamma$) field and quenching without resorting to mechanical or thermo-mechanical treatments. The initial alloy composition and volume pct. martensite are properly controlled so that the carbon concentration in the martensite phase is approximately 0.3 wt.%. As a result, during martensitic transformation the inhomogeneous shear component occurs by slip, not twinning (10). Hence, unlike many other composite systems in which the second phases are in most cases strong and brittle, the property of the martensite constituent will be both strong and tough, as has been shown in our parallel alloy design program on structural martensitic steels (10). Thus the results of this latter program have been helpful in the development of the duplex alloy steel program.

From a practical point of view, it is desirable to have a large slope in the A_3 line in order to have flexibility in heat-treatment for control of volume fraction of the two phases. This can be achieved by adding suitable ternary alloying elements (e.g., Si) (11) to the Fe-C base system, as is illustrated in Figure 1. In Figure 1a any slight variation in the two phase annealing temperature will shift composition as well as volume fraction of austenite (and hence martensite) to a great extent, whereas in the case of Figure 1b less stringent control on the critical annealing temperature is required, thereby increasing the flexibility for heat treatment.

It is also important to take into account the parameters such as optimum volume fraction of martensite (V_m), distribution of martensite particles, and minor constituents, e.g., carbides and retained austenite. Qualitatively, the lower limit of the acceptable range of V_m is restricted to the value where no strengthening of the duplex alloy occurs. The upper limit is set when the failure of the martensite immediately leads to the failure of the duplex structure. Of course the final range of V_m will be determined by particular specifications and applications for which the steel will be used. The system is obviously rather flexible.

In adding second phase particles to ductile matrices, ductility is usually sacrificed for strength. However, one can imagine from the many possible duplex distributions of martensite particles, a suitable morphology which should provide not only strengthening but also a minimal loss in ductility. In this regard, a fine scale, discontinuous, and fibrous distribution of martensite particles are considered to be desirable (4).

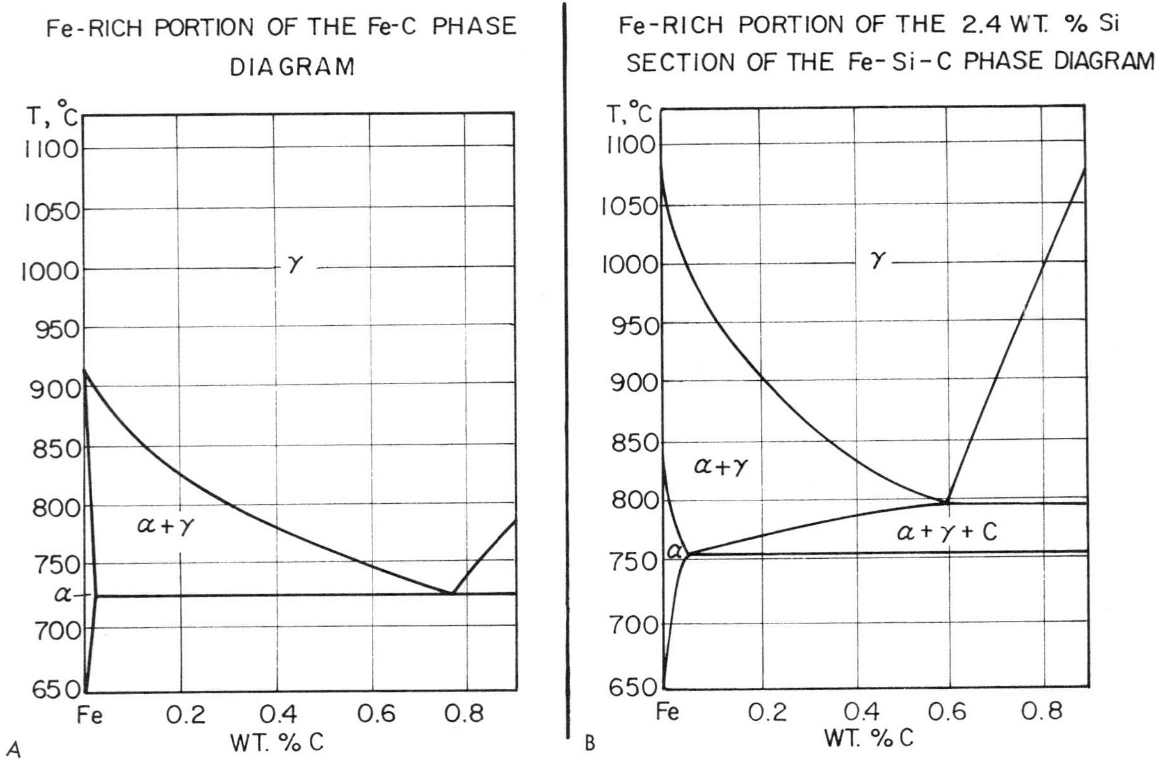

1. Phase diagrams showing the expansion of the (α+γ) range when silicon is added to the Fe-C system.

The composition of the DFM steels under consideration is based primarily on low carbon and low alloy. The resultant lack of sufficient hardenability may cause the ferrite-carbide (eutectoid) transformation products to be formed during cooling from austenite to martensite. These carbides, if any, will be located only in the immediate vicinity of the α/martensite interfaces, and therefore should be avoided (12). This can be accomplished only by quenching fast enough or by adding the proper alloying elements to suppress carbide formation.

The most difficult challenge is to produce DFM steels with these design considerations at a sufficiently small cost to make them economically competitive. We start with simple ternary Fe/X/C alloys prior to advancing to a more complex system (4). Metallurgical knowledge accumulated to date on the behavior of alloying elements (x) suggest that silicon is one of those which most favorably controls the design parameters of interest here (13). In addition, silicon provides very effective solid solution strengthening in the ferrite.

Experimental

The compositions of the alloys investigated are listed in Table I.

Table I. Alloy Composition (wt.%)

Alloy Number	Fe	C	Cr	Si	Mn
1010	bal.	0.10	--	--	0.5
1020	bal.	0.19	--	--	0.5
2	bal.	0.06	0.49	--	--
1	bal.	0.07	2.02	--	--
5	bal.	0.07	4.00	--	--
6	bal.	0.07	--	0.49	--
4	bal.	0.065	--	2.02	--

High purity 1010 and 1020 alloys were vacuum melted and cast as 20 lb ingots. These ingots were subsequently upset forged to approximately 0.5 in. thickness. Portions were further reduced to approximately 0.15 in. by cold rolling. The experimental Fe/X/0.1 C alloys were prepared and provided by Daido Steel Co. Japan. The alloys were melted in a vacuum induction furnace. The ingots were then forged in the range 1000-1100°C into 15 mm diameter rods. They were homogenized at 1200°C in vacuum for 20 hrs and subsequently furnace-cooled. The heat treatment to produce controlled DFM structures consisted of austenitizing and quenching to 100% martensite, followed by annealing in the (α+γ) range (details are described elsewhere (13)).

Tempering, when employed, was accomplished by the immersion of samples in a neutral salt bath kept at desired temperatures (200°C). All specimens were directly water quenched from the tempering temperature. Tensile properties were determined using 1 inch gage round tensile specimens, following ASTM specifications. Two inch gage round specimens were also used only when comparisons with the tensile properties of commercial steels were made. Tensile tests were performed at room temperature in an Instron machine with a cross-head speed of 0.05 cm/min and a full scale load of 1000 Kg.

The experimental procedures for optical metallography and transmission electron microscopy are given elsewhere (4).

Results and Discussion

DFM Structures Observed In Si Containing Steels:

Figure 2 shows an optical micrograph of the fibrous DFM structures developed in Si containing steels, as a result of the dual phase annealing and

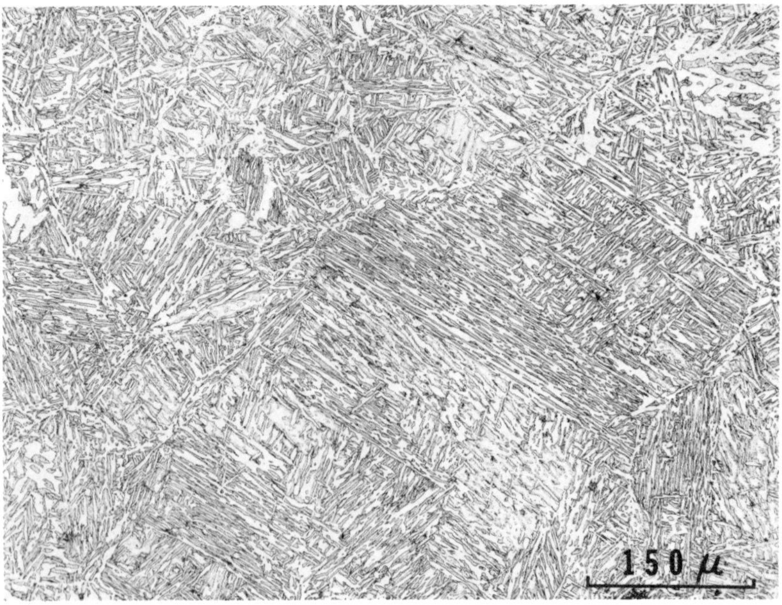

2. Optical micrograph of DFM structure developed in alloy 4 (Fe/2% Si/0.1 C).

quench treatment. A magnified view of the individual martensite particles is shown in the transmission electron micrograph (Figure 3), obtained from the 2% Si DFM steel. As the Si content was lowered, the major structural features remain unchanged but brittle carbides were present near the

3. Transmission electronmicrographs showing fibrous DFM structures developed in the 2 pct Si steel. Two parallel needles are martensite phase surrounded by ferrite with a high density of dislocations.

α/martensite interfaces. This is illustrated in Figure 4 taken from 0.5% Si DFM steel. This morphology of carbide distribution is very similar to that of the interphase precipitation transformation product observed in a variety of steels (14-15).

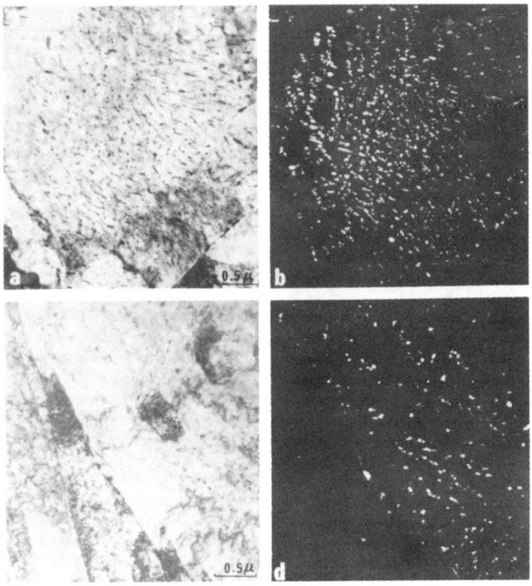

4. Morphology of heavy precipitation of carbides in the immediate vicinity of α/martensite boundaries. (a) and (c) Bright field, and (b) and (d) corresponding Dark field image of the carbides.

Nature of α/Martensite Interface:

Since the load bearing constituent is the martensite phase, the nature of the α-martensite interface (coherent or not) is of particular importance. Figure 5 shows the conventional bright field (a) and corresponding high resolution lattice fringe image (b) of an α/martensite interface in the 2% Si DFM steel. As the (110) fringes cross the interface, they are distorted but are continuous except for occasional end-on dislocations. The continuity in this case is interpreted as follows. Assuming the K-S orientation relationship holds in the $\alpha + \gamma$ mixture at 950°C then for $(111)_\gamma$ the particular variant of the six possible $(110)_\alpha$ will be the $(101)_\alpha$ which already exists parallel to $(111)_\gamma$ across the original $\gamma - \alpha$ interface. Hence on transformation $\gamma \rightarrow \alpha$, $(111)_\gamma$ becomes $(101)_\alpha$ parallel to $(101)_\alpha$ in the pre-existing ferrite. This is consistent with the measured d spacings.

It is known that strong particles that do not have good atomic fit with the matrix act as sites for failure by decohesion or by encouraging the formation of intercritical micro-cracks at the weak interface between particles and matrix. Ideally, the good coherency at the α/martensite interface revealed by lattice imaging will prevent such decohesive interface failure during deformation, and thus enable the full toughness of the ferrite to be realized. This high resolution electron micrograph not only provides striking information on the atomic arrangement near the interface, but also allows the determination of carbon concentration in the martensite particles. Any attempt to determine the carbon concentration from the equilibrium phase diagram will be inaccurate since the two phase annealing for 20 minutes is a non-equilibrium reaction. Analytical measurements involving x-ray and electron probe techniques become extremely difficult for the submicron martensite particles (e.g., Figure 3). Lattice fringe imaging, however, is a powerful technique in analyzing extremely localized chemical composition.

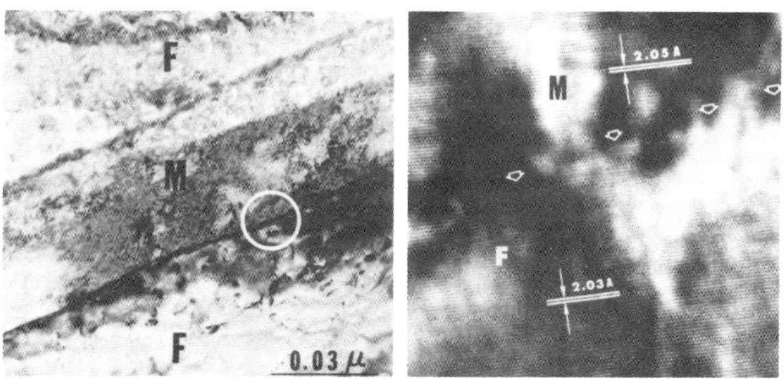

5. Conventional bright field (a) and lattice image electron microscopy (b) of a α/martensite interface in the 2% Si DFM steel. The lattice image (b) was taken from the area encircled in (a). Martensite tetragonality creates the larger d_{101} spacing in the martensite region (M). "F"-ferrite. The arrows indicate the interface.

Tensile Properties

The room temperature tensile test data of the as-quenched DFM steels are plotted in Figures 6 and 7. Figure 6 shows the variation of yield and tensile strengths of the steels as a function of martensite volume fraction. The composite strength obeys the rule of two phase mixtures in the range of 15∿80 pct martensite, irrespective of the composition and morphology. The variation of uniform and total elongation with respect to pct martensite are plotted in Figure 7, which demonstrates the validity of the two phase mixture rule for elongation in the range of 10∿80% martensite. From these plots, it is clear that the 2% Si DFM steels exhibit the best combinations of strength and elongation over a wide range of pct martensite. The 0.2 pct offset yield and ultimate strengths of the duplex 2% Si steel as a function of total elongation are shown in Figure 8, compared with some selected commercial HSLA steels including Van 80.

The attractive features of the 2% Si DFM steel are perhaps better illustrated in the characteristic stress-strain behavior (Figure 9). The relatively low yield strength coupled with extremely high rate of work hardening in the early stage of plastic deformation of the 2% Si DFM alloys 4S1 (20% MS), 4S2 (40% MS), 4S3 (60% MS) results in high tensile/yield strength ratio and good elongation to necking, as is shown in Figure 9. Also shown in Figure 9 for comparison are the engineering stress-engineering strain curves for a fully martensitic structure (4A) and a commercial HSLA steel, Van 80 (16).

In the early stage of plastic deformation of the composite, the strong phase particles (martensite) do not deform with the soft phase matrix (ferrite). Mobile dislocations are impeded by the particles. The dislocation density increases rapidly as further deformation proceeds, resulting in an increase in effective size of the particles. In this way, stresses are built up, in and around the particles. The resultant stress-strain features include early and fairly extreme work hardening rates which do suppress mechanical instabilities. As a consequence, the transition from elastic to plastic deformation is smooth without showing yield point phenomena that have been observed in a number of metals and alloys both in the single and polycrystalline state, e.g., note the yield point elongation in the load-elongation curve for Van 80. This mode of behavior continues until the yield stress of the martensite particle is reached at about 3% deformation. Here dislocations will start to cut through the semi-coherent martensite particles on the slip systems that are common to ferrite and martensite. As deformation proceeds the load required for further deformation increases. However, after the onset of large scale yielding the work hardening rate of the DFM alloys decreases and becomes virtually equal to that of Van 80, regardless of the martensite volume fraction.

The gradual equalization of work hardening rates at high strains is in agreement with calculations using finite element method (17), which indicated that when the hard phase (martensite or carbide) yields, the work hardening rate of the composite becomes equal to that of ferrite but at a higher flow stress level.

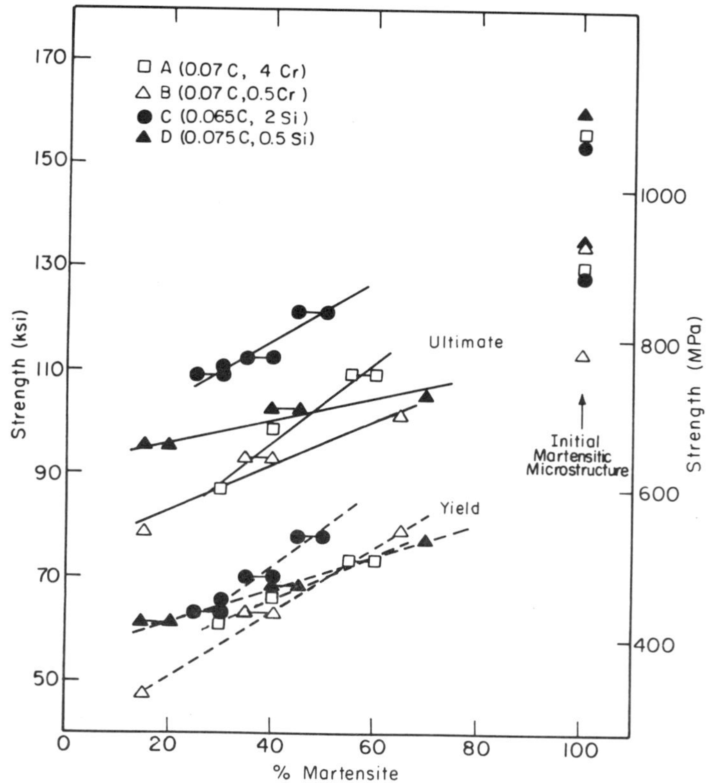

6. Yield and ultimate tensile strengths as a function of martensite volume fraction for DFM steels.

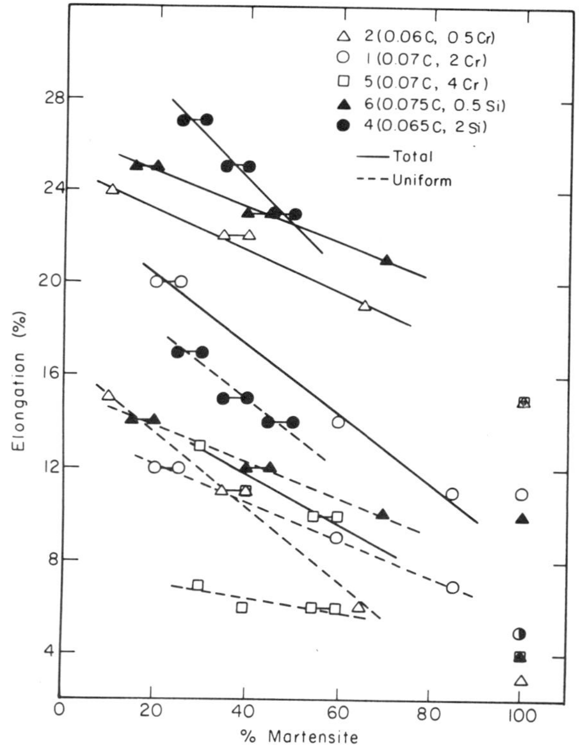

7. Uniform and total elongations as a function of martensite volume fraction for the DFM steels.

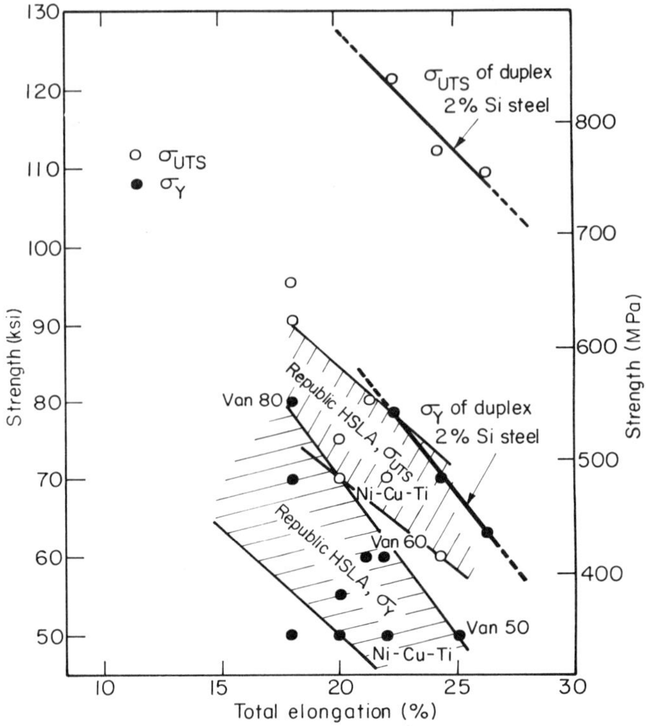

8. Tensile properties of the duplex 2 pct Si steel are compared with those of commercial HSLA steels.

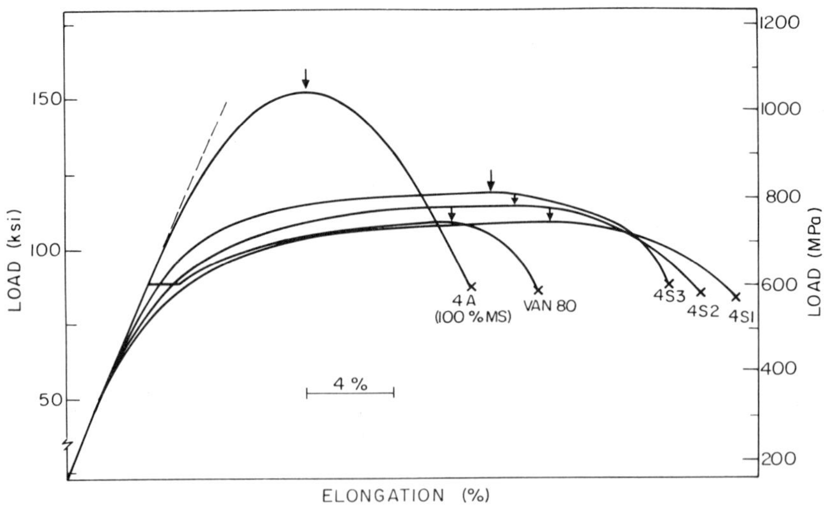

9. Load-elongation (engineering stress-engineering strain) curves for the 2 pct Si DFM alloy with varying amount of martensite volume fraction (20,40, and 60%), specimen 4A having 100% martensite, and Van 80.

Influence of Ferrite Purity and Martensite Strength:

It is important to understand how the properties of a DFM steel can be affected as those of the constituent phases are changed. Tempering experiments on the 2% Si DFM steel (containing ∿30% martensite) have been conducted to consider the case where the ferrite properties are varied while the martensite properties are kept constant.

Tempering at 200° C for 1 hr did not cause any appreciable change in the morphology and mechanical properties of the martensite phase in the DFM steel. This was confirmed by transmission electron microscopy and isolated experiments measuring the mechanical properties of 100% martensite structure before and after 200° tempering. However, the ferrite regions which were free from any detectable precipitation in the as-quenched condition are now associated with carbide precipitation, as is seen in Figure 10. The coarse carbides, ∿170 Å wide and ∿1500 Å long, were identified as cementite. The other areas in the ferrite where these coarse carbides were not present showed finely dispersed precipitation on dislocation networks, at the nodes of intersecting dislocations and at subboundaries (Figure 11). Such a morphological change in the ferrite can be correlated with the significant increase in the elongation of the tempered DFM steel without appreciable loss in strength, as shown in Table II.

Table II. Tensile Test Summary (Tempered 2% Si DFM structure*)

Specimen	Tempering Temp. (°C)	σ_y (Ksi)	σ_{UTS} (Ksi)	Unif. elong. (%)	Total elong. (%)	R.A. (%)
4S	AQ**	65	110	16	23	59
4ST1	200	59	103	17	30	70

* At 30% martensite.
** As quenched.

Upon tempering, the supersaturated carbon atoms in the ferrite matrix of the as-quenched DFM steel are depleted from the solution to improve ductility of the ferrite, thereby giving rise to the sharp increase in the total elongation of the DFM steel.

Now consider the case where the strength of martensite is varied while maintaining the other metallurgical variables constant. For this purpose, 1010 and 1020 steels were duplex treated to yield identical morphology of DFM structures, the only difference being the carbon concentration (higher in DFM 1020) in the martensite at a given volume pct. The results are shown in Figure 12, where σ_y, σ_{UTS}, and uniform elongation are not significantly affected but total elongation was decreased as the martensite strength decreased. This result is consistent with the design principles described previously. The above considerations indicate that the ferrite matrix must be highly ductile and martensite constituent strong and tough to achieve the best balance of strength and ductility of the composite at a given pct martensite.

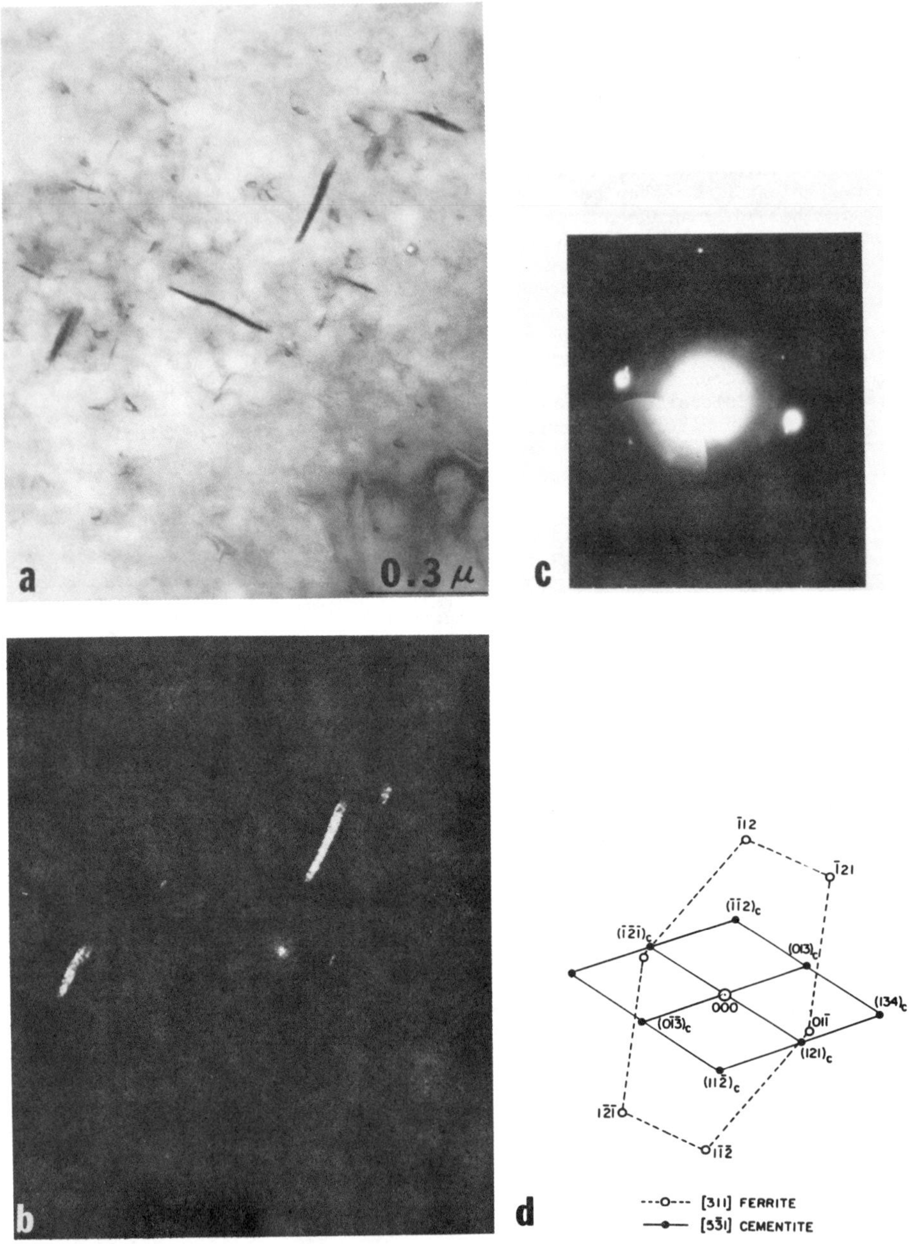

10. (a) Bright field and (b) dark field electron micrographs showing precipitation of coarse carbides in a ferrite region of the 0.5 Cr DFM steel tempered at 200°C for 1 hour. (c) SAD of (a). (d) The coarse carbides, ~170 Å wide and ~1500 Å long were identified as cementite by indexing diffraction pattern.

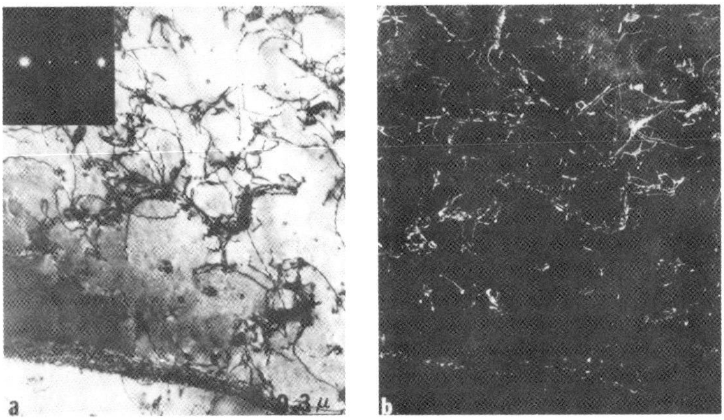

11. Transmission electron micrograph of a ferrite region in the 2% Si DFM steel after 200°C tempering for 1 hour, showing finely dispersed precipitation on dislocations and at sub-boundaries. (a) Bright field image does not clearly resolve the fine scale precipitates. However, weak beam image (b) shows greatly improved resolution of the heterogeneous precipitation on dislocations and at sub-boundaries. (g, 3g) weak beam imaging condition was used ((a) inset).

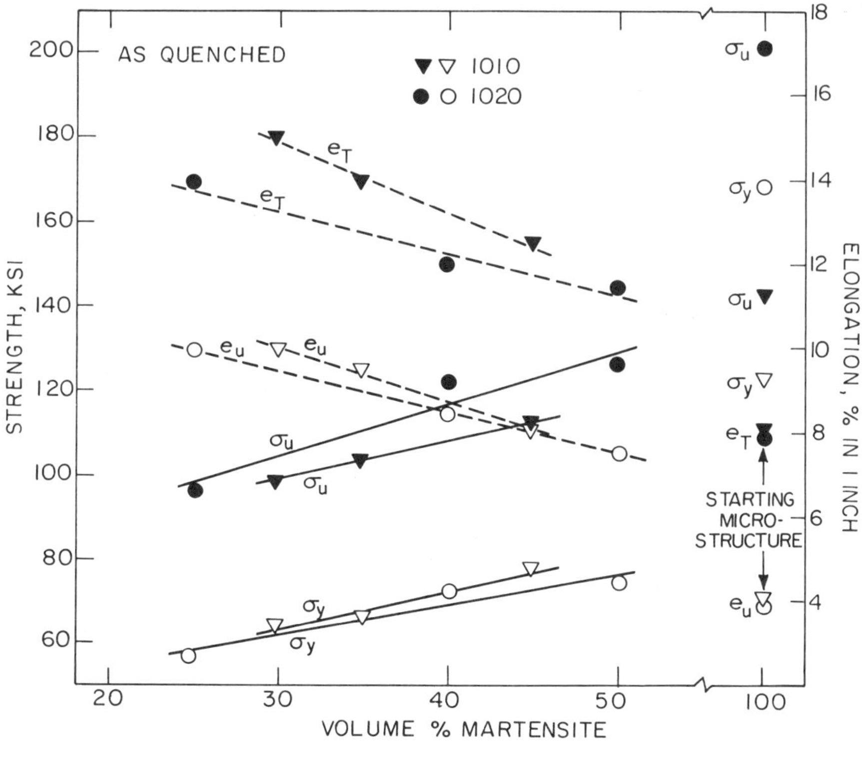

12. Strengths and elongation ductility vs. martensite fraction for DFM 1010 and 1020 steels.

Conclusions

1. Design principles for improved mechanical properties of DFM steels have been suggested in order to obtain desirable microstructural characteristics which in turn result in desirable mechanical properties.

2. An alloy, of composition Fe/2% Si/0.1% C, has been selected according to the design principles.

3. The 2% Si DFM steel showed high strength and good formability, e.g., $\sigma_y \sim 70$ Ksi, $\sigma_{UTS} \sim 112$ Ksi, uniform elongation $\sim 15\%$, and total elongation $\sim 25\%$ at about 40% martensite. These properties are superior to a series of Cr and Mn containing DFM steels, and to some commercial HSLA steels including Van 80.

4. Over the range of $15 \sim 80\%$ martensite, the rule for two phase mixtures appears to hold as a fairly good approximation of the tensile behavior of the duplex systems.

5. Slight improvements in the ductility of ferrite appear to cause a significant increase in the elongation ductility of DFM structures.

6. The total elongation of DFM structures is sensitive to the carbon content in the martensite phase whereas σ_y, σ_{UTS}, and uniform elongation are relatively insensitive.

7. Lattice fringe imaging of the α/martensite interface revealed that (110) fringes were continuous across the interface, indicating coherency is maintained.

Acknowledgments

We would like to thank Dr. Asada and Daido Steel Company for supplying the experimental alloys used in this research. Appreciation is also extended to M.J. Young for providing some experimental data.

This work was done with support from the U.S. Department of Energy through the Materials and Molecular Research Division of Lawrence Berkeley Laboratory, University of California, Berkeley.

References

1. Nakaoka, K., Araki, K., and Kimpara, O., Sosei to Kako, Vol. 17, No. 190, 1976, p. 875.

2. Hayami, S., and Furukawa, T., Micro Alloying 75, Symposium on High Strength, Low-Alloy Steels, Products and Process, Washington, D.C., 1975, p. 56.

3. Rashid, M.S., SAE Preprint 760206, Feb. 1976.

4. Koo, J.Y., and Thomas, G., Mat. Sci. and Engin., Vol. 24, 1976, p. 187. see also, Koo, J.Y., Ph.D. Thesis, University of California, Berkeley, LBL-6657, Aug. (1977).

5. Grabe, A.G., U.S. Patent 2097878, (1937).

6. Herres, S.A., and Lorig, C.H., Trans. ASM, Vol. 40, 1945, p. 775.

7. Morita, S. et al., U.S. Patent 3288657 (1966).

8. Cairns, R.L., and Charles, J.A., JISI, Vol. 205, 1967, p. 1044.

9. Grange, R.A., Proceedings of the 2nd Int'l Conf. on the Strength of Metals and Alloys, ASM, Pacific Grove, 1970, p. 861.

10. Thomas, G., Battelle Conf. on the Fundamental Aspects of Structural Alloy Design, Richland, Washington, Jaffee, R.J., and Wilcox, B.A., eds., Plenum Press, 1977, p. 331-361.

11. Piwowarsky, E., Hochwertiges Gußeisen, Spring-Verlag, 1951, p. 52.

12. Mogford, I.L., Met. Rev., Review 114, Vol. 12, 1967, p. 49.

13. Koo, J.Y., and Thomas, G., Met. Trans., Vol. 8A, 1977, p. 525.

14. Davenport, A.T., Berry, F.G., and Honeycombe, R.W.K., Metal. Sci. J., Vol. 2, 1968, p. 104.

15. Tamino, M., and Aoki, K., Trans. Iron Steel Inst. Japan, Vol. 8, 1968, p. 339.

16. Davis, R.G., Ford Research Report #SR-77-29, 1977.

17. Fishchmeister, H., Hjalmered, J.-O., Karlsson, B., Linden, G., and Sundstrom, B., "The Microstructure and Design of Alloys", Proc. 3rd Int'l Conf. on Strength of Metals and Alloys, Inst. of Metals & Iron Steel Inst., London, 1973, p. 621.

LABORATORY STUDIES OF MICROSTRUCTURES AND STRUCTURE-PROPERTY RELATIONSHIPS IN "DUAL-PHASE" HSLA STEELS

J. M. Rigsbee and P. J. VanderArend*

Republic Steel Corporation
Research Center
Independence, Ohio

Introduction

Over the past several years, the necessity for increasing the miles-per-gallon (MPG) efficiency of automobiles has prompted extensive industry-wide automotive size and mass (weight) reduction programs. With vehicle downsizing approaching the limits of acceptability for the average size family and with increasingly stringent average fleet fuel economy government regulations, weight reduction by means of direct materials substitution is a major portion of the automotive industry's general plan for meeting MPG requirements. An economical way to achieve significant weight reductions is through gage reductions made possible by the substitution of currently available high-strength low-alloy (HSLA) steels for the presently used lower strength plain carbon steels. The limiting factor in this approach is the inherently lower formabilities that available HSLA steels have in comparison to the formabilities of plain carbon steels. "Dual-phase"** steels, which should become commercially available industry-wide within the next five years, offer much improved formabilities at strengths equivalent to current HSLA steels (1, 2). This class of steels appears to be a viable material for achieving weight reduction in automobiles through substitution for plain carbon steels (1).

During the last decade there have been a number of investigations (1-12) of steels having a ferrite matrix and containing a less deformable "second-phase" (usually termed martensite). Although Cairns and Charles (3, 4) found work hardening to increase in ferrite + martensite microstructures, they also found that addition of small volume fractions of martensite to ferrite resulted in a large ductility decrease. This ductility decrease is almost certainly due to the synthetic nature of their lamellar ferrite/martensite composites. Grange (5) later showed that creation of martensite in a ferrite matrix through annealing resulted in much reduced yield strength/ultimate tensile strength ratios and considerably improved elongations. A following investigation by Matsuoka and Yamamori (6) showed that intercritical annealing, producing a ferrite + martensite microstructure, resulted in a decreased

---

* Now at the University of Chicago, School of Business.

** "Dual-phase" refers to the steel's microstructure, which typically is reported to consist of ferrite + martensite.

yield strength, continuous yielding, a high rate of work hardening, and a virtually unchanged ultimate tensile strength. Hayami and Furukawa (7) also showed in their study of "dual-phase" steels that superior YS/UTS ratios, continuous yielding, and improved elongations can be obtained. Similar results have recently been published by Rashid (1, 8), Dabkowski and Speich (9), Baily and Stevenson (10), Araki et al. (11) and Davies (12). Rashid (1, 8) has shown the formability of "dual-phase" steels to be superior to conventional HSLA steels with equal UTS values and has established the following "dual-phase" steel mechanical property requirements:

- No yield point elongation (a smooth, continuous stress-strain curve);
- A 0.2% offset yield strength of 345 ± 39.5 MPa (50 ± 5 ksi);
- At 2 to 4% strain, a flow stress of 520-550 MPa (75-80 ksi);
- An ultimate tensile strength of 620-655 MPa (90-95 ksi);
- A total elongation of ≧ 27%*.

The objective of this investigation was to study the microstructures and mechanical properties developed by isothermally annealing two HSLA steels at temperatures ranging from 732°C (1350°F) to 954°C (1750°F). Although the annealing process is not commercially feasible and the annealed materials usually do not develop the complete range of mechanical properties required to be classified as a "dual-phase" steel, the range of mechanical properties and range of microstructures developed by the annealing treatments allowed "dual-phase" steel structure-property correlations to be obtained. Quantitative analysis of the microstructural constituents was accomplished using optical (light) microscopy, scanning electron microscopy (SEM), transmission electron microscopy (TEM), and x-ray diffraction techniques.

Experimental Techniques

Material and Heat Treatment

The compositions (wt %) of the two steels used in this investigation are given in Table I. These steels are similar in composition to MAXI-FORM** 80, a commercial HSLA steel with a 550 MPa (80 ksi) yield strength. Prior to heat treatment, the microstructure of these as-hot-rolled steels consisted of fine grained, equiaxed ferrite (ASTM 13-14) with grain boundary carbides, a very low volume fraction of pearlite and a dispersion of fine vanadium and columbium carbonitride precipitates. The materials were sectioned transversely into tensile blanks and ground to a uniform final thickness of 4.45 mm (0.175 in.).

Table I. Alloy Compositions

Alloy	C	Mn	Si	Cb	V	Cr	Mo	Ni	Al	N
I	0.072	1.26	0.30	0.077	0.08	0.02	< 0.01	0.02	0.035	0.008
II	0.11	1.29	0.29	0.071	0.065	0.03	< 0.01	0.02	0.038	0.008

* In a 2.5 mm (0.1 in.) thick tensile sample with a 51 mm (2.0 in.) long by 12.5 mm (0.5 in.) wide gauge section.

** MAXI-FORM is a registered trademark of Republic Steel Corporation.

The tensile blanks were heat treated as follows:

1. A 15-minute preheat in a circulated-air furnace at 316°C (600°F) to remove any moisture.

2. Immediately transferred to and held for 5 minutes in a neutral salt pot at temperatures ranging from 732°C (1350°F) to 954°C (1750°F) in 55°C (100°F) increments.

3. Immediately quenched in heated oil (49°C (120°F)).

Tensile Testing

After heat treatment, the blanks were machined into tensile samples with gauge sections 51 mm long by 12.5 mm wide. The mechanical properties were evaluated on an Instron tensile testing machine using a crosshead speed of 0.508 mm (0.2 in.) per minute to failure. Yield strength (0.2% offset) and ultimate tensile strength (UTS) values were determined in the conventional manner and uniform elongations were measured at the maximum load on the load-elongation charts. The rate of work hardening was determined for each tensile sample as the average work hardening parameter, i.e., n value (13), for the region of the load-elongation curve between 1 and 5% strain. Duplicate tensile samples were tested for each material/heat treatment combination.

Microstructural Analysis

Optical and Scanning Electron Microscopy: Through-thickness sections from the tensile sample grip ends were examined using standard optical metallography and scanning electron microscopy (SEM) techniques. Using an etch consisting of 2 gms $(NH_4)_2S_2O_8$, 2 ml HF, 50 ml CH_3COOH and 150 ml H_2O, it is possible to distinguish the ferrite, retained austenite and martensite microconstituents by the differing degrees of etching attack.

Transmission Electron Microscopy (TEM): Slices ca. 0.4 mm (0.016 in.) thick were obtained from the non-gripped portion of the tensile sample grip ends with a low-deformation sample sectioning device (Buehler Isomet) and were thinned to ca. 0.1 mm (0.004 in.) using the chemical polishing solution described by Plichta, Aaronson and Lange (14). To avoid introduction of dislocations and possible strain-induced transformation of the retained austenite microconstituent through accidental deformation of a thin foil, 3 mm (0.125 in.) diameter discs were produced from the thinned sheets using a photoetching technique (15). The discs were then "dished", to produce a thinned center region supported by a thicker outside rim, with a single jet polishing device (16) using an 8 vol % perchloric acid in glacial acetic acid solution at room temperature and at a potential of ca. 250 volts. The profiled discs were thinned to perforation in a 250 ml glacial acetic acid, 75 gm anhydrous sodium chromate, 25 gm chromic oxide and 10 ml water solution at < 10°C and ca. 28 volts polishing potential. This solution was found to thin ferrite, retained austenite and martensite at virtually identical rates. Most of the TEM studies were done with the Siemens ELMISKOP 102 located in the Department of Metallurgy and Materials Science at Case Western Reserve University.

Quantitative Metallography: Multiphase stereology relationships developed by Cribb (17) were used for quantitative analysis of the ferritic matrix and "second-phase" retained austenite and martensite microconstituents for each material/heat treatment combination. Standard point counting and line intercept measurements (18) were used with the developed relationships to determine volume fractions, mean particle diameters and mean free path in the ferrite matrix between particles for both the retained austenite and martensite. Because of the difficulty in distinguishing (in an optical micrograph at 2000X) subtle etching differences* between "second-phase" martensite and retained austenite particles ca. 1 micron or less in diameter, the retained austenite volume fraction determined for each sample with x-ray analysis techniques was considered more accurate than the stereology value and was therefore used for structure-property correlations. Since the total "second-phase" volume fraction determined for each sample through the stereology relationships was correct and consisted of only the individual retained austenite and martensite volume fractions, the actual martensite volume fraction present in the microstructure and used for structure-property correlations was equal to the total "second-phase" volume fraction determined with stereology minus the x-ray determined retained austenite volume fraction.

X-Ray Analysis: Measurement of retained austenite volume fractions with x-ray techniques followed a standardized procedure in which ratios of the areas under austenite and ferrite peaks are converted, with the aid of standards and charts, into volume fractions (19). Since many of the retained austenite particles are less than 10^{-3} mm in diameter, it is felt that x-ray analysis gives a more accurate determination of volume fraction. X-ray analysis also has the advantage of averaging over a much larger area than is practical with the stereology technique. To ensure a deformation free surface for x-ray analysis**, surface grinding with 600 grit paper was followed by chemically polishing (13) to a depth of ca. 0.1 mm (0.004 in.) from the as-ground surface. Retained austenite measurements were obtained at two different through-thickness levels in the plane of the tensile strip.

Results

Mechanical Properties

As shown by the listing of mechanical properties in Table II, the additional heat treatments produced significant changes in the strengths, elongations and work hardening characteristics of the two steels relative to their as-hot-rolled properties. Figure 1, a plot of 0.2% offset yield strength*** (OYS) and ultimate tensile strength (UTS) as a function of annealing temperature, shows that the annealing treatments affect the strengths of each material similarly. Except for the small, gradual increase in UTS found for the higher carbon material, UTS appears to be relatively insensitive to annealing temperature. However, annealing temperature does have a major influence on OYS, with each material obtaining a minimum OYS after annealing at 843°C (1550°F). Annealing at temperatures above 843°C was found to cause the OYS to increase, with the higher carbon material increasing most rapidly. In comparison to the as-hot-rolled condition, annealing each material at 843°C for 5 minutes results in an OYS decrease of ca. 240 MPa (35 ksi) and an OYS/UTS ratio decrease from ca. 0.87 to 0.55.

---

* As will later be discussed in more detail, the etchant used in this investigation was found to etch martensite more than retained austenite.

** As will later be shown, the retained austenite is unstable with respect to mechanical deformation.

*** The yield strength of materials having less than ca. 0.5% Lüders strain was determined with the conventional 0.2% offset yield point technique.

Table II. Summary of As-Hot-Rolled and Heat Treated Sample Mechanical Properties

Thermal Treatment	Carbon Level	0.2% Yield Strength, MPa (ksi)	Ultimate Tensile Strength, MPa (ksi)	% Lüders Strain	Work Hardening Parameter, n	% Uniform Elongation	% Total Elongation
As-Hot-Rolled	0.072	638 (92.6)	718 (104.0)	2.2	0.12	12.0	22.0
		640 (92.7)	726 (105.0)	2.2	0.12	12.0	22.0
		639 (92.7)	722 (104.5)	2.2	0.12	12.0	22.0
As-Hot-Rolled	0.11	620 (89.9)	723 (104.3)	2.4	0.13	12.0	23.5
		621 (90.0)	721 (104.5)	2.6	0.13	12.5	22.5
		621 (90.0)	722 (104.7)	2.5	0.13	12.3	23.0
5 Min at 732°C (1350°F), OQ	0.072	547 (79.3)	743 (107.7)	0.9	0.159	16.0	23.5
		523 (75.8)	749 (108.6)	0.7	0.167	15.0	22.0
		535 (77.6)	746 (108.2)	0.8	0.163	15.5	22.8
5 Min at 732°C (1350°F), OQ	0.11	499 (72.3)	752 (109.0)	0.7	0.187	15.0	22.5
		495 (71.8)	754 (109.2)	0.7	0.195	16.0	24.0
		497 (72.1)	753 (109.1)	0.7	0.191	15.5	23.3
5 Min at 788°C (1450°F), OQ	0.072	434 (62.9)	709 (102.7)	0.7	0.222	14.0	22.5
5 Min at 788°C (1450°F), OQ	0.11	485 (70.4)	743 (107.7)	0.4	0.203	15.0	19.0
		418 (60.5)	841 (121.9)	0	0.238	15.0	19.0
		445 (64.5)	792 (114.8)	0.2	0.220	15.0	19.0
5 Min at 843°C (1550°F), OQ	0.072	380 (55.1)	684 (99.2)	0	0.253	20.0	25.5
		375 (54.3)	679 (98.4)	0	0.252	19.0	24.5
		377 (54.7)	682 (98.8)	0	0.253	19.5	25.0
5 Min at 843°C (1550°F), OQ	0.11	387 (56.1)	768 (111.3)	0	0.253	18.0	24.0
		403 (58.4)	778 (112.8)	0	0.258	18.5	23.5
		395 (57.3)	773 (112.1)	0	0.256	18.3	23.8
5 Min at 899°C (1650°F), OQ	0.072	419 (60.7)	705 (102.2)	0	0.206	20.0	25.5
		413 (59.9)	696 (100.9)	0	0.207	20.0	26.5
		416 (60.3)	701 (101.6)	0	0.207	20.0	26.0
5 Min at 899°C (1650°F), OQ	0.11	502 (72.7)	816 (118.2)	0	0.191	17.0	23.0
		501 (72.6)	824 (119.4)	0	0.203	15.0	21.0
		502 (72.7)	820 (118.8)	0	0.197	16.0	22.0
5 Min at 954°C (1750°F), OQ	0.072	451 (65.4)	715 (103.6)	0	0.181	21.1	28.0
		434 (62.9)	707 (102.5)	0	0.176	22.5	26.5
		443 (64.2)	711 (103.1)	0	0.179	21.8	27.3
5 Min at 954°C (1750°F), OQ	0.11	565 (81.9)	842 (122.0)	0	0.183	11.0	19.0
		558 (80.8)	835 (121.0)	0	0.217	13.0	19.0
		561 (81.4)	838 (121.5)	0	0.200	12.0	19.0

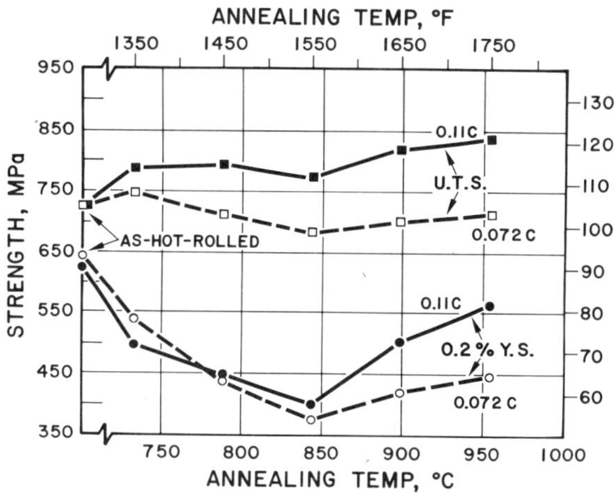

Figure 1. Variation in yield strength and ultimate tensile strength for both materials as a function of annealing temperature.

Figure 2 is a plot of uniform and total elongations for each material as a function of annealing temperature. At 843°C and below, the effects of annealing temperature on the elongations of each material appear to be similar. In comparing elongation values of as-hot-rolled material and annealed material at temperatures up to 843°C, the major effect of the heat treatment is an increase of ca. 7% in the uniform elongation, with the total elongation increasing only ca. 2%. For annealing temperatures above 843°C, elongation changes relative to the 843°C values appear to be related to carbon content, with uniform and total elongations both increasing for the lower

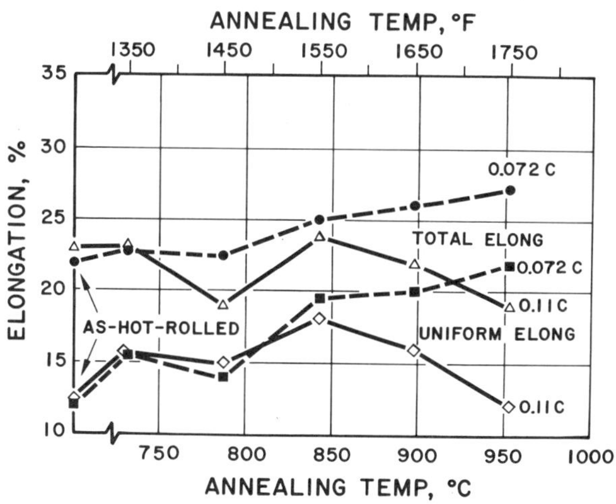

Figure 2. Variation in uniform and total elongation for both materials as a function of annealing temperature.

carbon material and both decreasing for the higher carbon material. In comparing the as-hot-rolled material and annealed material elongations, it is clear that the annealing treatments reduce the necking elongations (i.e., the amount of elongation to fracture following attainment of the UTS) of both materials.

In Figure 3, uniform and total elongations are plotted as a function of UTS for each material in both the as-hot-rolled and heat treated conditions. This plot shows that improved uniform and total elongations can be achieved by decreasing the UTS of the material. As a group, the lower carbon material has a lower average UTS and a higher average elongation than the higher carbon material. It is apparent from this figure that the major elongation difference between as-hot-rolled and heat treated materials with similar UTS values occurs in the uniform elongation. It is also apparent from this figure that the necking elongation for the heat treated material is about one-half the value for the as-hot-rolled material.

Figure 4 is a plot of work hardening parameter (n) versus annealing temperature and shows that the effects of annealing temperature on the work hardening characteristics of each material are nearly identical. The "error-bars" on the individual points in this figure represent the data spread (see Table II). The n value initially increases with increasing annealing temperature, reaches a maximum of ca. 0.25 at 843°C (1550°F) and decreases with higher annealing temperatures. Depending on annealing temperature, the ratio of n values for annealed versus as-hot-rolled materials ranges from ca. 1.4 to 2. Later discussions will explain the relationship between the improved uniform elongations obtained with the annealed materials (Figures 2 and 3) and the greater work hardening capacities of the annealed versus the as-hot-rolled materials (Figure 4).

Microstructural Analysis

Optical and Scanning Electron Microscopy: Figure 5 is a series of optical micrographs showing the microstructure of the lower carbon steel in both

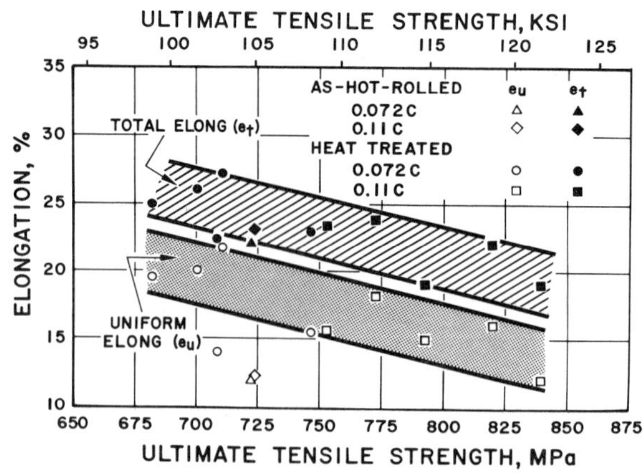

Figure 3. Correlation between elongation (uniform and total) and ultimate tensile strength for both materials and all annealing temperatures.

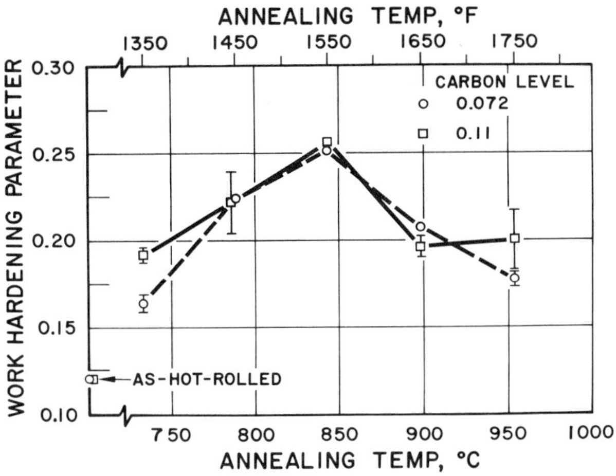

Figure 4. Variation in work hardening for both materials as a function of annealing temperature.

the as-hot-rolled and the heat treated conditions*. The as-hot-rolled microstructure in Figure 5(a) shows elongated ferrite grains (ca. ASTM 13 to 14) with a low volume fraction of small grain boundary carbides. Following the 732°C (1350°F) heat treatment, Figure 5(b) shows that the ferrite grains remain elongated and that the grain boundary carbides have been replaced by a new "second-phase" microconstituent present as small particles nucleated on the ferrite grain boundaries. After the 788°C (1450°F) anneal, Figure 5(c) shows that the ferrite grains have become equiaxed** and that the grain boundary nucleated "second-phase" particles have increased in both average size and volume fraction. After the 843°C (1550°F) anneal, Figure 5(d), the ferrite grains appear little changed from the 788°C anneal. The grain boundary nucleated microconstituent is somewhat changed, with some of the larger "second-phase" particles etching darker than the remaining smaller particles. Annealing at 899°C (1650°F), Figure 5(e), causes the ferrite grains to increase in size and causes the volume fraction and average particle size of the "second-phase" to noticably increase***. It is now apparent that the "second-phase" is actually composed of two distinctly different microconstituents, one that is present as large particles and etches darker than the ferrite matrix and a second that is present as small particles and etches lighter than the ferrite matrix. After annealing at 954°C (1750°F), the microstructure shown in Figure 5(f) is very similar to that produced by the 899°C anneal. The matrix ferrite (F), large "second-phase" particles (M) and small "second-phase" particles (A) are identified in Figure 5(f). As

---

* The as-hot-rolled and the heat treated microstructures of the 0.11 C steel were equivalent to those of the 0.072 C steel.

** At this temperature and higher, ferrite grain boundaries are not strongly attacked by the etchant and appear indistinctly on the micrographs.

*** Quantitative analysis of volume fractions and particle sizes for the "second-phase" in both materials and for all heat treatments is given later in this paper.

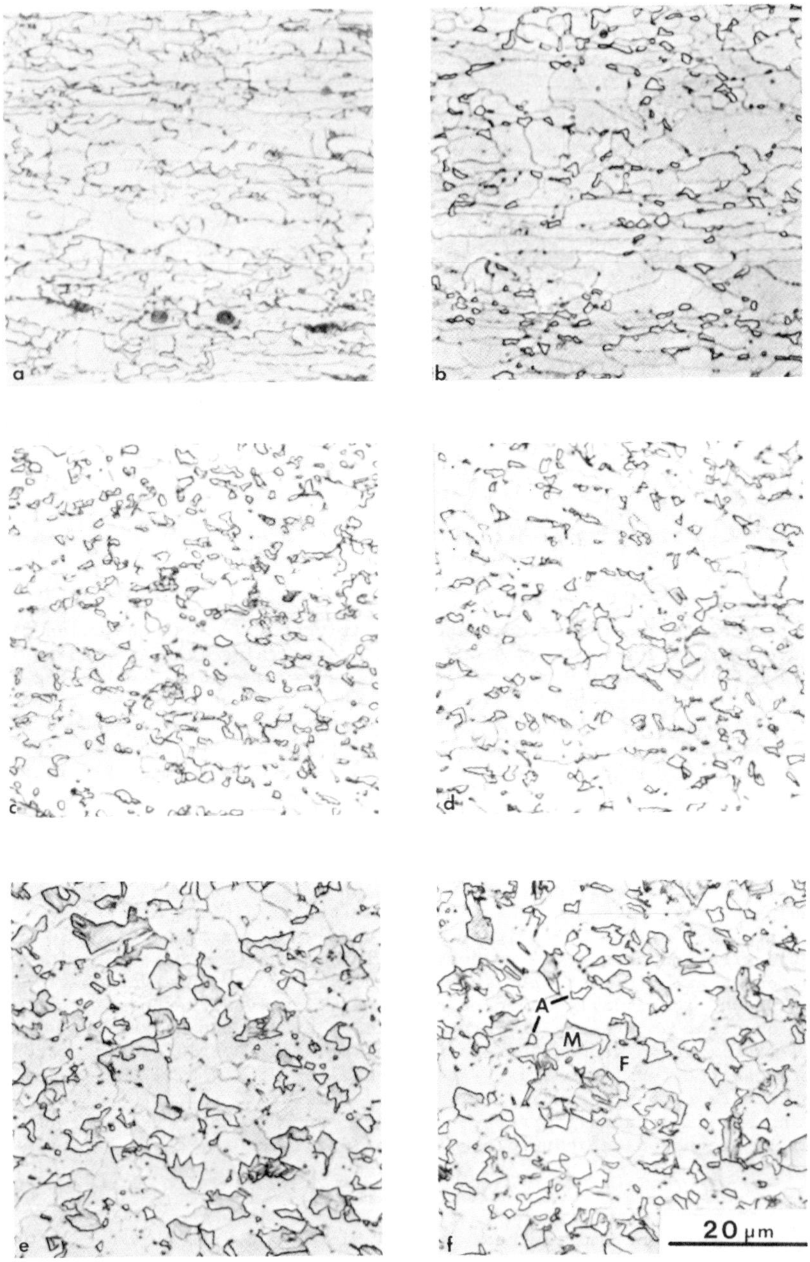

Figure 5. Microstructural changes in the 0.072 wt % C steel at various annealing temperatures: (a) as-hot-rolled; (b) 732°C (1350°F); (c) 788°C (1450°F); (d) 843°C (1550°F); (e) 899°C (1650°F); and (f) 954°C (1750°F).

later will be proven with x-ray and TEM diffraction and quantitative metallography techniques, the large "second-phase" particles are martensite and the small particles are retained austenite. The retained austenite and martensite normally exist as separate particles, although small untransformed austenite regions are sometimes found associated with larger martensite particles. Almost all of the martensite and retained austenite particles are derived from prior austenite regions which appear to have nucleated on the ferrite grain boundaries (frequently at triple points) during the heat treatment.

High magnification SEM examination of etched samples revealed the source of the etching contrast differences between the retained austenite and martensite particles. The retained austenite particles shown in Figure 6(a) have smoother surfaces than the ferrite and, because of their higher surface reflectivity, would therefore appear lighter in an optical micrograph than the ferrite background. Conversely, the larger martensite particle shown in Figure 6(b) has a rougher surface than the ferrite and would appear darker in an optical micrograph than the ferrite background. The very different etching characteristics of the retained austenite and martensite microconstituents are most likely a result of the internal structure differences (discussed later in more detail). The general lack of internal structure for the retained austenite and the profuse amount of internal structure for the martensite would favor the martensite being most rapidly attacked by the etchant.

Transmission Electron Microscopy: Figure 7 consists of low and medium magnification TEM micrographs showing ferrite grains and "second-phase" particles for the 0.072 C material heat treated at 899°C (1650°F). In the low magnification micrograph, Figure 7(a), the equiaxed ferrite grains appear as the lighter microconstituent and are ca. 3 to 4 microns in diameter. The "second-phase" particles appear as the very dark microconstituent and are ca. 1 to 2 microns in diameter. From the locations of the "second-phase" particles, the austenite formed during the heat treatment process appears to have nucleated preferentially on the ferrite grain boundaries, with the triple points being the most active nucleation sites. Figure 7(b), a higher

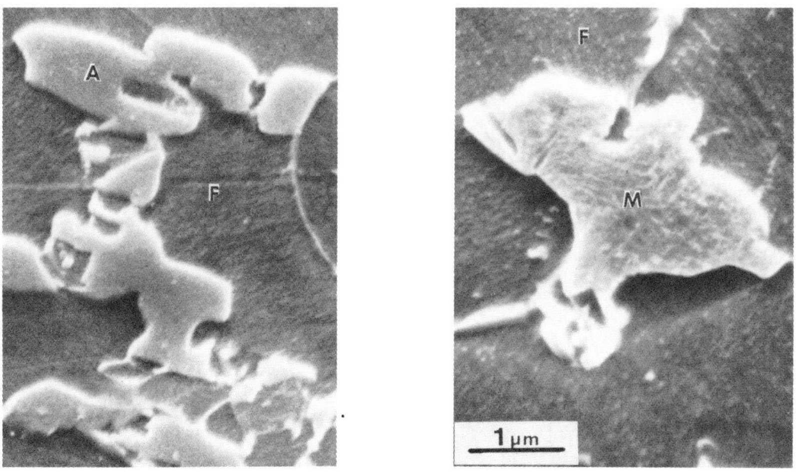

(a) (b)

Figure 6. SEM examination of "second-phase" microconstituents: (a) retained austenite and (b) martensite.

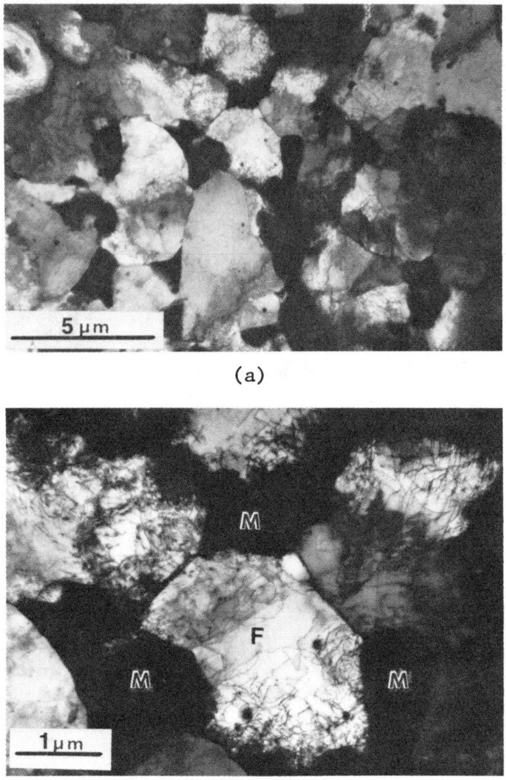

(a)

(b)

Figure 7. TEM bright-field micrographs showing equiaxed ferrite grains (F) and "second-phase" particles (M): (a) low magnification and (b) higher magnification.

magnification view of the upper central region of Figure 7(a), shows the dislocation structure within the ferrite grains. Most of the dislocation structure is concentrated in the ferrite regions directly adjacent to the "second-phase" particles, with the central portions of the ferrite grains having a relatively low dislocation density. Since the "second-phase" particles appear to be the source of the ferrite dislocation structure and since dislocations have been shown to be generated by the volume change accompanying the austenite to martensite transformation (7, 8, 20-22), the particles are identified as martensite.

Figure 8 consists of bright-field and dark-field TEM micrographs which show a higher resolution example of the dislocation distribution in the ferrite regions near a martensite particle. The particle occupying the lower left-hand quarter of Figure 8(a) was identified as martensite by its internally twinned microstructure (23, 24). The bright-field micrograph, Figure 8(a), clearly shows that the dislocation structure generated by the austenite to martensite transformation is concentrated in the ferrite region directly adjacent to the martensite particle. The dark-field lower magnification micrograph, Figure 8(b), shows a low dislocation density in the ferrite regions 1 micron or more away from the martensite particle.

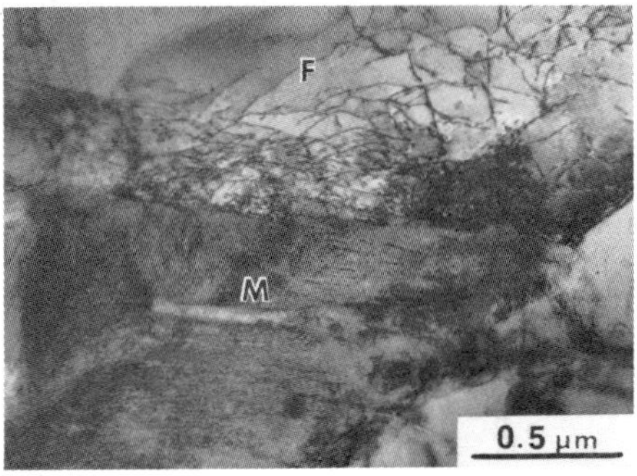

(a)

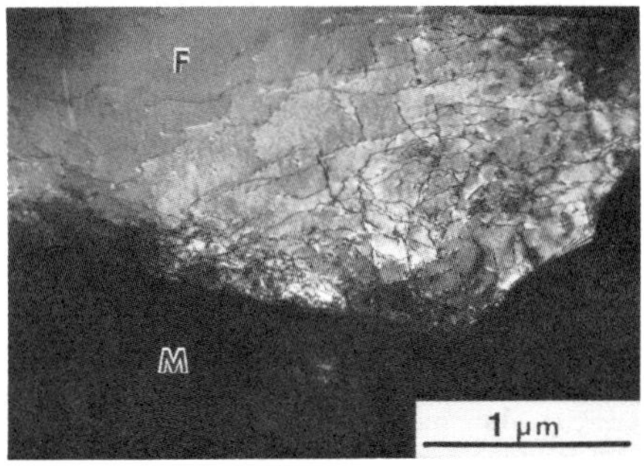

(b)

Figure 8. TEM examination of variation in dislocation density in ferrite (F) adjacent to a martensite (M) particle: (a) bright-field and (b) dark-field.

(a)

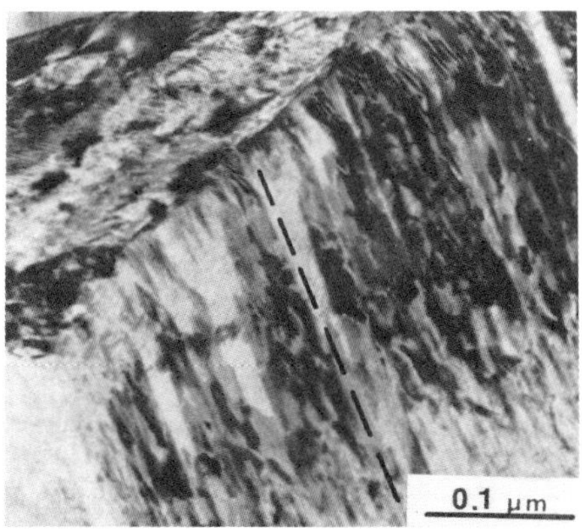

(b)

Figure 9. TEM bright-field micrographs showing a martensite particle containing several variants: (a) low magnification and (b) high magnification showing fine, internally twinned microstructure.

Figure 9(a) and 9(b) show low and high magnification views of a typical martensite particle. As evident from Figure 9(a), this martensite particle is composed of a complex arrangement of several transformation variants. The higher magnification TEM micrograph shown in Figure 9(b) illustrates the internal transformation twin structure of two adjacent martensite variants. All martensite regions observed in this investigation exhibited the internally twinned structures characteristic of high carbon, low transformation temperature plate martensite (23, 24) and most martensite regions contained more than one martensite transformation variant. No lath martensite, a lower carbon and higher transformation temperature form of martensite, was observed.

As earlier stated, in addition to the darker etching martensite microconstituent, a very small (typically submicron sized) lighter etching microconstituent exists and is retained austenite. Figure 10 shows an example of a martensite particle with a small appendage of retained austenite. The crystal structure of this small appendage was determined with selected area electron diffraction. Since the carbon contents of the large martensite particle and its appendage should be very similar, this is a good example of size stabilization of the retained austenite particles.

Figure 11 shows a detailed TEM analysis of a retained austenite particle. Figure 11(a) is a bright-field micrograph showing the ferrite matrix and the retained austenite particle. In Figure 11(b), the retained austenite particle is imaged in dark-field with a $<111>_{f.c.c.}$ austenite diffraction vector. The numerous stacking faults present in the retained austenite particle are the probable nucleation sites for the austenite to martensite phase transformation. The availability in the austenite of four different $\{111\}_{f.c.c.}$ type fault planes could explain the origin of the multiple martensite variants typically found in a single prior austenite particle. The indexed electron diffraction patterns shown in Figures 11(c) and 11(d) correspond respectively to the austenite and ferrite and provide quantitative crystal structure identification.

Since the effect of a second phase on altering the mechanical properties of the matrix is related to the mutual deformation constraints (i.e., the compatability) at the second phase/matrix interface (25, 26), the coherency of the austenite/ferrite interface was investigated. In Figure 12, the interfaces between the austenite particle and the ferrite matrix exhibit well-

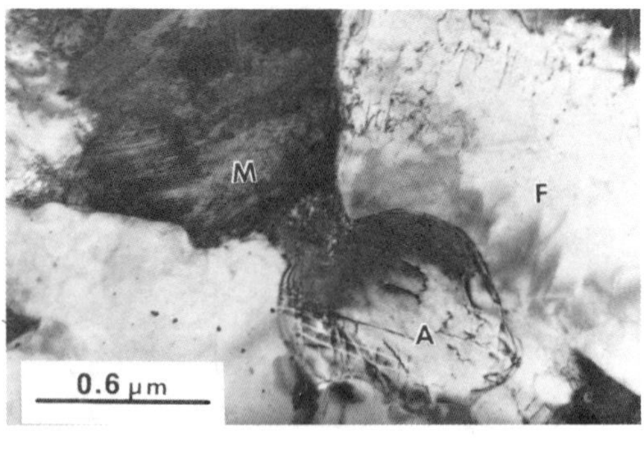

Figure 10. TEM bright-field micrograph showing an austenite region which has partially transformed to martensite: F is ferrite, M is martensite, and A is retained austenite.

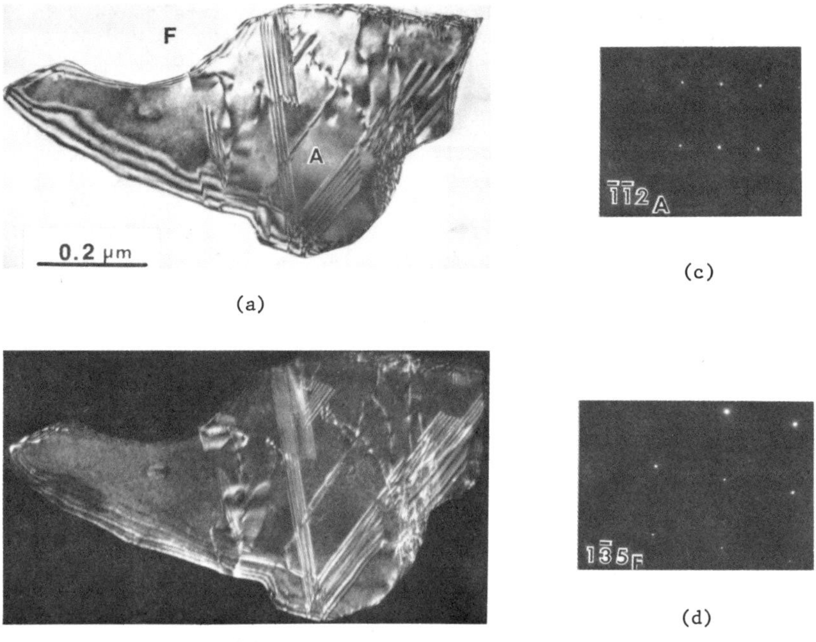

Figure 11. TEM analysis of a retained austenite particle (A) lying in a ferrite matrix (F): (a) bright-field micrograph; (b) dark-field micrograph; (c) selected area diffraction pattern from the retained austenite particle; and (d) diffraction pattern from the ferrite.

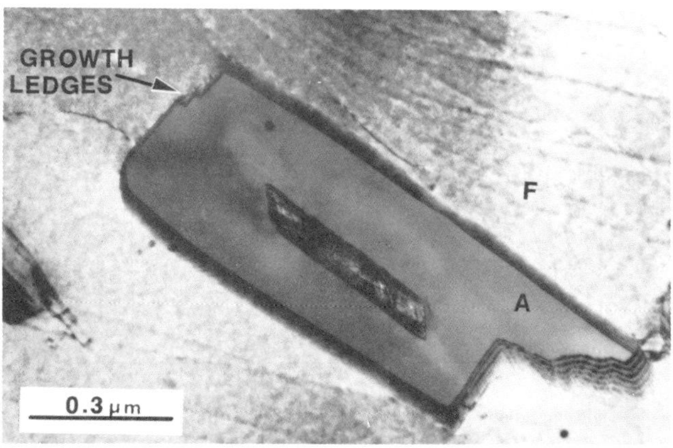

Figure 12. TEM micrograph of retained austenite particle (A) lying in a ferrite matrix (F). The faceted A/F interfaces and the labeled growth ledges indicate partial coherency between the phases.

defined faceting (for this case, a facet is a planar interphase interface). The interfacial faceting indicates the existance of a structural barrier to growth (27), which means that the austenite/ferrite interface is at least partially coherent. The 150 Å to 200 Å high growth ledges labeled in Figure 12 give additional proof that the austenite/ferrite interfaces are partially coherent. Well-defined faceting was frequently observed, although it was not found for every austenite particle examined. The conclusion that austenite/ferrite interfaces are normally semicoherent is consistent with the recent results of other investigations involving structural modeling of f.c.c.:b.c.c. interfaces (28-30) and TEM analysis of f.c.c.:b.c.c. interfacial structures (31).

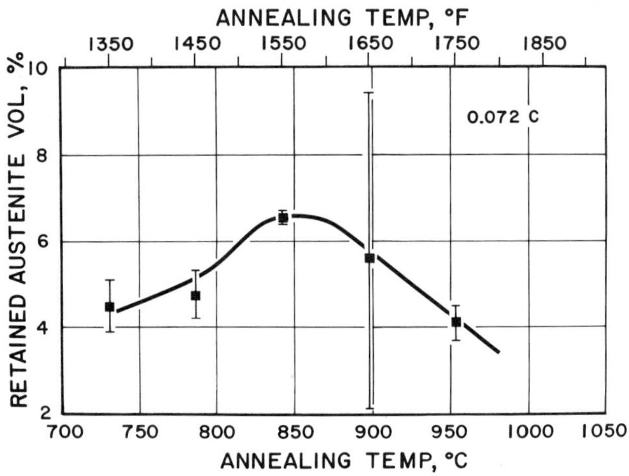

(a)

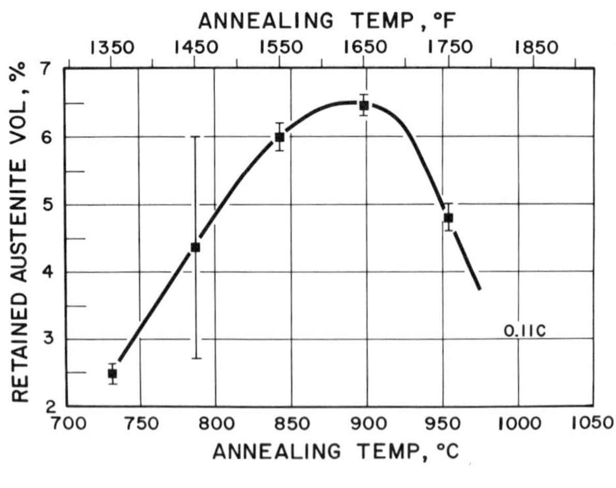

(b)

Figure 13. X-ray analysis of retained austenite volume fraction as a function of annealing temperature: (a) 0.072 C material and (b) 0.11 C material.

X-Ray Diffraction (XRD): Figures 13(a) and 13(b) show the results of volume fraction retained austenite XRD analyses for both materials as a function of annealing temperature. The amount of retained austenite in each material initially increases with increasing annealing temperature, reaches a maximum of ca. 6.5 volume percent for annealing temperatures around 843°C (1550°F) to 899°C (1650°F) and decreases for higher annealing temperatures. The bars on the data points correspond to the data spread.

In an attempt to determine the austenite to martensite transformation temperature (i.e., M_s), samples were immersed for times of up to 10 minutes in liquid nitrogen (-196°C) prior to XRD analysis. No decrease in the volume fraction of retained austenite was noted and tensile samples immersed in liquid nitrogen showed no change in mechanical properties when pulled at room temperature. Since it has been observed by A. R. Marder that the retained austenite in a similar "dual-phase" material is stable even down to liquid helium (-269°C) temperature (32), the retained austenite has no practical M_s temperature.

XRD analysis of the lower carbon material annealed at 843°C and strained in tension at room temperature showed that the retained austenite undergoes a deformation induced transformation to martensite (i.e., $M_D > RT$). Figure 14 shows a plot of the variation in the retained austenite volume fraction as a function of tensile strain (bars on the data points correspond to the data spread). During the first five percent strain slightly less than 50% of the retained austenite transforms to martensite. The rate of loss of retained austenite per unit strain decreases with increasing strain and nearly 20% of the retained austenite remains untransformed at strains near the uniform elongation (19.5%).

Quantitative Metallography: The results of the quantitative microstructural analysis of the "second-phase" retained austenite and martensite microconstituents are shown in Table III for both materials and for all annealing temperatures. Data on volume fraction, mean particle diameter and mean free

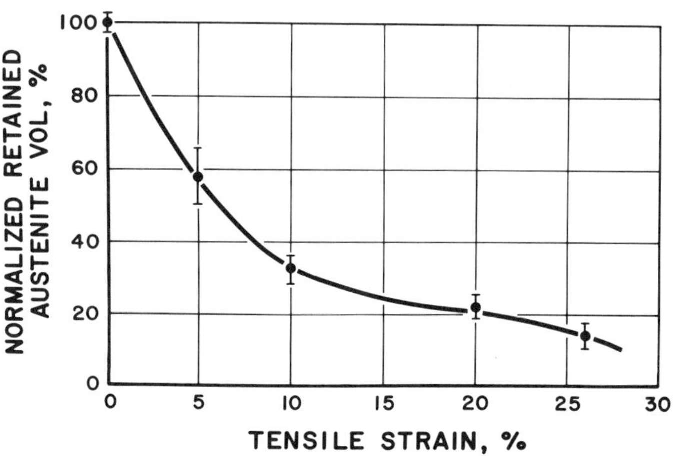

Figure 14. X-ray analysis of normalized retained austenite volume fraction as a function of tensile strain.

Table III. Summary of Quantitative Metallographic and X-Ray Analyses of Heat Treated Sample Microstructures

		Austenite				Martensite				Total Second Phase	
Alloy	Anneal Temperature, °C	Quantitative $V_V^A \pm \sigma$	X-Ray V_V^A	$\bar{\lambda}^A$	$\bar{\lambda}_A^\alpha$	Quantitative $V_V^M \pm \sigma$	Adjusted $V_V^M \pm \sigma$	$\bar{\lambda}^M$	$\bar{\lambda}_M^\alpha$	$V_V^A + V_V^M$	$\bar{\lambda}_{A+M}^\alpha$
I	732	5.79 ± 1.84	4.5	0.75	12.0	0.73 ± 0.43	2.12 ± 0.84	1.83	233	6.62 ± 2.27	11.4
I	788	6.2 ± 3.05	4.8	0.65	9.7	0.73 ± 0.81	2.13 ± 1.5	0.77	98	6.93 ± 3.86	8.8
I	843	10.3 ± 3.97	6.6	0.88	7.6	0.73 ± 0.59	4.43 ± 2.02	0.71	87	11.03 ± 4.56	7.0
I	899	3.3 ± 0.83	5.8	0.73	19.1	10.0 ± 2.86	7.5 ± 2.15	2.38	20	13.3 ± 3.6	9.9
I	954	9.09 ± 1.44	4.1	1.3	11.9	7.64 ± 1.0	12.63 ± 1.79	2.08	23	16.73 ± 2.44	7.8
II	732	4.39 ± 1.59	2.5	0.38	8.3	0	1.84 ± 0.68	—	—	4.39 ± 1.59	8.3
II	788	7.44 ± 1.44	4.3	0.60	7.5	0.11 ± 0.35	3.25 ± 0.96	0.28	232	7.55 ± 1.79	7.3
II	843	7.96 ± 2.34	6.0	0.65	7.4	1.45 ± 0.85	3.41 ± 1.43	1.4	88	9.41 ± 3.19	6.8
II	899	6.4 ± 2.05	6.4	0.66	9.4	3.1 ± 0.95	3.1 ± 0.95	1.5	44	9.5 ± 3.0	7.7
II	954	9.5 ± 2.89	4.8	0.91	8.4	2.81 ± 1.21	7.51 ± 2.64	7.2	77	12.4 ± 4.1	7.6

V_V^i = Volume percent of ith phase

$\bar{\lambda}^i$ = Mean particle diameter of ith phase (mm x 10^{-3})

$\bar{\lambda}_i^\alpha$ = Mean free path in ferrite matrix between particles of ith phase (mm x 10^{-3})

path in the ferrite matrix between particles is given for each microconstituent. The x-ray retained austenite volume fractions and adjusted martensite volume fractions* are the volume fraction data used for correlation with mechanical properties. The differences between the x-ray and quantitative retained austenite volume fraction values for a given sample and annealing temperature are caused by the difficulty in distinguishing optically the etching characteristics of sub-micron sized particles. The basic results from this quantitative microstructural analysis are:

1. The retained austenite particles are, on the average, smaller than the martensite particles. The average of the mean particle diameters for both materials and for all annealing temperatures shows a mean particle size of 0.75 microns for the retained austenite and 2.0 microns for the martensite.

2. The retained austenite particles are more numerous and more closely spaced than the martensite particles. Even considering the previously discussed existance of some small martensite particles mistakenly classified as retained austenite, the mean interparticle spacing for martensite is still a factor of two to five times larger than the mean retained austenite interparticle spacing.

3. The adjusted martensite volume fraction does not begin to increase rapidly until the high end of the annealing temperature range is reached. Because this sharp increase is accompanied by a decrease in the retained austenite volume fraction, the total "second-phase" volume fraction increases at a nearly constant rate over the entire annealing temperature range investigated.

Microstructure-Mechanical Property Correlations

Figure 15**, a plot of yield strength versus martensite volume fraction for both materials at all annealing temperatures, shows the effect of martensite on altering yield strength. Introduction of as little as 2 volume percent martensite, through the lowest temperature annealing treatment, reduces the yield strength of both materials ca. 90 MPa (13 ksi) below their as-hot-rolled values. The yield strength initially decreases with increasing martensite content, reaches a minimum value of ca. 385 MPa (56 ksi) at ca. 4 to 5 volume percent martensite and increases with further increases in martensite content. Beyond the optimum martensite content for minimum yield strength, the rate of increase in yield strength with increasing martensite content appears to be related to the carbon content of the material (note the position of point 5 for the 0.11 C material in Figure 15).

As previously shown in Figure 4, the work hardening capacities of both materials following annealing were greatly improved in comparison to their as-hot-rolled values and varied rather widely for the range of annealing temperatures investigated. Comparison of the work hardening parameter with

---

* The adjusted martensite volume fraction is equal to the total "second-phase" volume fraction determined with the stereology-quantitative metallography relationships minus the x-ray retained austenite volume fraction.

** The numbers assigned to each data point in this and following figures correspond to the annealing temperature (i.e., 1 = 732°C, 2 = 788°C, 3 = 843°C, 4 = 899°C and 5 = 954°C.

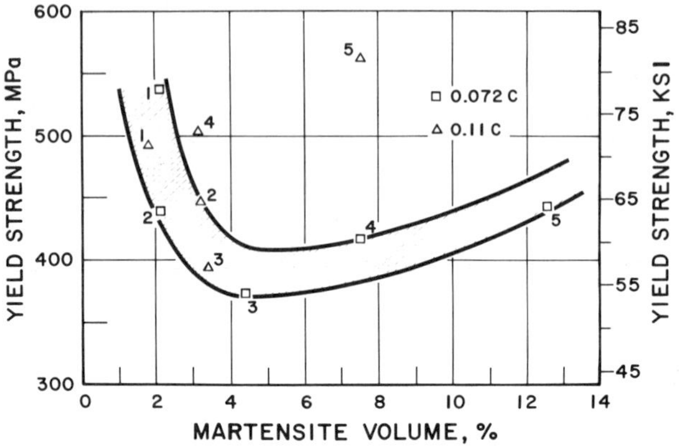

Figure 15. Variation in yield strength with increasing martensite volume fraction.

the total "second-phase" volume fraction, Figure 16, shows that the work hardening parameter increases with increasing "second-phase" content up to ca. 10 volume percent and decreases with further increases in total "second-phase" content. Since the higher "second-phase" volume fractions occur for the higher annealing temperatures where martensite is the largest volume fraction component of the "second-phase", it appears that martensite is not the major cause for increased work hardening in the 1% to 5% strain region. The plot of work hardening parameter versus retained austenite, Figure 17, does show a consistent increase in work hardening parameter with increasing retained austenite over the entire annealing temperature range investigated. This figure shows clearly that materials with the highest work hardening parameters are those which contain the largest volume fraction of retained austenite and do not contain more than ca. 5 volume percent martensite.

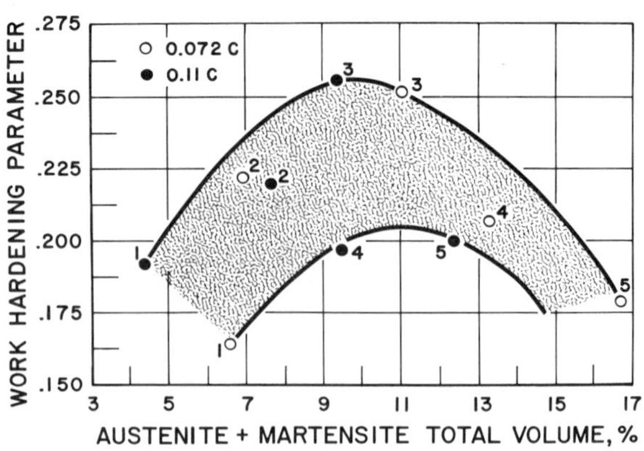

Figure 16. Variation in work hardening parameter with increasing volume fraction "second-phase".

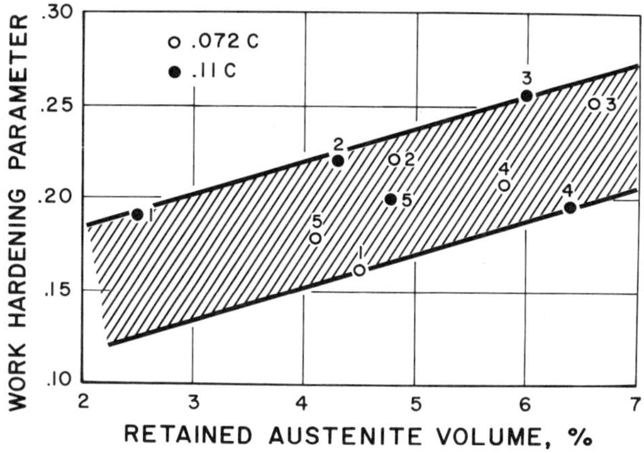

Figure 17. Variation in work hardening parameter with increasing retained austenite volume fraction.

Further evidence of the role of retained austenite is given by the plot of work hardening parameter versus mean austenite interparticle spacing, Figure 18. As is to be expected if retained austenite plays the key role in work hardening for the 1% to 5% strain regime, the highest work hardening parameters correspond to the instances where the retained austenite particles are most closely spaced.

As illustrated in Figure 3, when considering materials with similar UTS values the annealing treatments cause little change in the total elongations of as-hot-rolled versus annealed materials. However, the uniform elongation of the annealed material is increased by ca. 6% over the value for the as-hot-rolled material with a similar UTS. Figure 19, a plot of uniform elongation versus retained austenite volume percent, shows the cor-

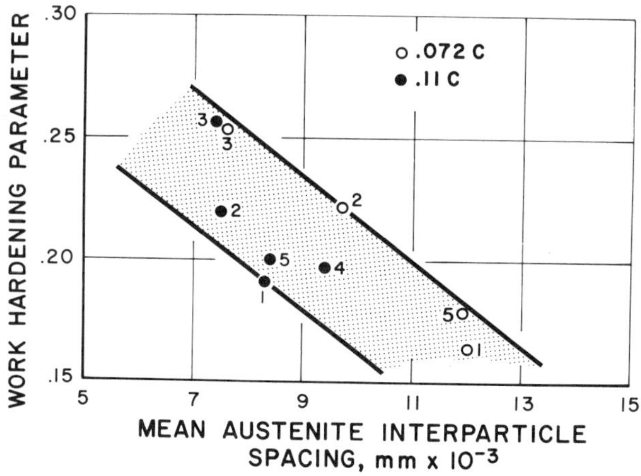

Figure 18. Correlation between work hardening parameter and mean austenite interparticle spacing.

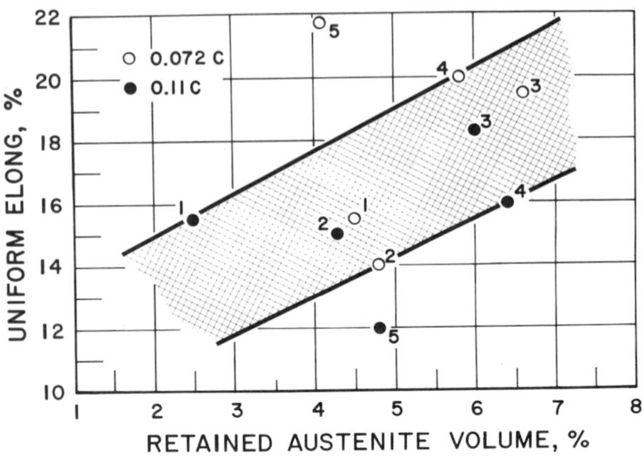

Figure 19. Variation in uniform elongation with increasing retained austenite volume fraction.

relation between increasing uniform elongation and increasing retained austenite volume percent. Detailed discussion of this phenomena is later presented and is based on the effects of retained austenite on the work hardening characteristics of the materials.

Discussion of Results

Microstructural Analysis

The annealing treatments employed in this study developed "dual-phase" microstructures in both the low and higher carbon steels over the entire temperature range investigated. Ferrite matrix grains, which were mostly elongated in the as-hot-rolled materials, became equiaxed when annealed at 788°C (1450°F) or higher. Annealing also eliminated the small grain boundary carbides in the as-hot-rolled materials and created austenite particles which appeared to nucleate almost totally on the ferrite grain boundaries, e.g., see Figures 5(b) and 7(a). It is expected that the austenite particles nucleate primarily at the grain boundary carbides since the driving force* for austenite nucleation is highest at this site (33) because of the greater ferrite carbon concentration and since the high carbide/ferrite interphase interface energy would reduce the surface free energy barrier for formation of an austenite nucleus (34). The amount of "second-phase" austenite formed during annealing and the mean austenite particle size both increased with increasing annealing temperature.

Upon cooling, those austenite particles larger than ca. 1 micron normally transformed completely to martensite and those smaller than ca. 1 micron remained austenitic. It is important to note that the "second-phase" produced by the annealing treatments in this investigation consisted of discrete particles of martensite and of retained austenite. The nature of this "second-phase" is considerably different from the "second-phase" M-A

* As given by the free energy change for the ferrite to austenite phase transformation.

microconstituent discussed by Biss and Cryderman (35) and Dabkowski and Speich (9) where the particles normally contained both martensite and retained austenite and from the strictly martensitic "second-phase" reported by a number of previous investigators of "dual-phase" materials (1, 6, 7, 10, 12).

Martensite: The martensite particles were typically 1 to 3 microns in size, were uniformly distributed throughout the material and contained little or no retained austenite. Although TEM analysis showed that the internal martensite structure invariably contained the fine twins characteristic of a plate martensite transformation mechanism (23, 24), the multiple martensite variants normally found in a single particle, e.g., Figure 9(a), almost never exhibited the well-defined lenticular shape usually associated with plate martensite. This shape deficiency is attributed to size constraints placed upon the martensitic transformation by the relatively small volume of austenite in which the transformation is confined.

Retained Austenite: The smaller particles of austenite formed during annealing were stabilized by their size and/or chemistry and remained austenitic even when cooled to -269°C (32). Although carbon and manganese are known to stabilize austenite by depressing the M_s temperature (36), complete partitioning of carbon and significant partitioning of manganese* to the austenite could not produce an M_s temperature below 0°C. It is suggested that an extra increment of stability is imparted to the retained austenite particles by their extremely small size. A similar size stabilization effect has been noted by Kinsman et al. (20) for extremely small (1000 Å diameter) iron rich f.c.c. precipitates in a copper matrix and by Leslie and Miller (37) in an Fe-Ni alloy with an extremely fine grain size (ASTM 19).

Although it is not rigorously understood how the austenite to martensite transformation is inhibited by a small austenite particle size, the mechanism will likely involve both martensite nucleation and growth. As demonstrated experimentally by Cech and Turnbull (38) and later discussed by Magee (39), when a fixed volume of austenite is divided into discrete, disconnected regions it is possible to isolate regions which do not contain an effective martensite nucleation site. It is also possible that even when an effective martensite nucleus is present (e.g., the stacking faults in the austenite particle shown in Figure 11) the austenite particle can be stabilized by the matrix constraints resulting from the partially coherent nature of the ferrite/austenite interface (e.g., Figure 12).

Both of these concepts are compatible with the results of this investigation which show the retained austenite to be mechanically unstable when deformed at room temperature. Deformation of the material would introduce deformation structure into the austenite particles which could serve as a martensite nucleus. Deformation of the material would also produce strain gradients in the ferrite at the ferrite/austenite interface. These strain gradients could effectively reduce the constraints of the partially coherent interface by reducing the amount of ferrite deformation required to maintain compatability across the interface during the shape change accompanying the transformation of an austenite particle to martensite (40).

---

* Significant manganese enrichment of the austenite is not expected due to the slow diffusion rate of manganese in ferrite and due to the relatively low temperatures and short times used for heat treatments in this investigation.

Microstructure-Mechanical Property Correlations

The range of annealing temperatures employed in this investigation produced a wide range of microstructures and mechanical properties in both steels. Although the heat treated materials do not completely satisfy the "dual-phase" GM980X mechanical property specifications listed in the Introduction, correlations between the microstructures and mechanical properties for these materials provide a basis for understanding the microstructural origin of the unusually good combination of strength and formability exhibited by "dual-phase" steels.

Yield Strength and Continuous Yielding: GM980X mechanical property specifications state that there should be no yield point elongation (i.e., a smooth, continuous stress-strain curve) and that the 0.2% yield strength should be 345 ± 35 MPa (50 ± 5 ksi). These specifications require that regions suitable for initiation of slip under a low applied stress be uniformly distributed throughout the material. Comparison of yield strength versus annealing temperature (Figure 1) and microstructure versus annealing temperature (Table III) shows a good correlation between decreasing yield strength and increasing martensite volume fraction (Figure 15). As an explanation of the relationship between martensite volume fraction and yield strength, it has been shown that the austenite to martensite transformation is accompanied by a volume expansion of ca. 3% to 4% (41-43) and that this volume expansion can create mobile dislocations in the matrix regions immediately adjacent to the martensite regions (7, 8, 20-22). Since the martensitic transformation of the austenite patches occurs at relatively low temperatures (due to the high carbon content of the particles), the dislocations generated in the ferrite will not be solute pinned and are thus mobile.

TEM examination of heat treated material (e.g., Figures 7 and 8) showed that the austenite which transformed to martensite produced heavily dislocated, "pre-yielded" ferrite regions which lie in close proximity to the martensite particles. Since the martensite particles are uniformly distributed throughout the material and since the ferrite grain interiors have low dislocation densities and do not contain appreciable amounts of fine microalloy carbonitride precipitates which could serve as a barrier to slip, motion of the already present and mobile transformation induced dislocations can occur at a low level of stress and can explain the low yield strength. Residual stresses generated in the ferrite by the volume expansion accompanying the austenite to martensite transformation are another contributing factor in the martensite-yield strength relationship (9, 22, 44). Because the transformation occurs at low temperatures, the transformation induced stresses will be only partially relieved by the generation of dislocations in the ferrite and the residual stresses can assist in the initiation of yielding at low stress levels.

A martensite volume fraction of ca. 4% to 5% is found to produce the minimum yield strength for both materials. The increase in yield strength noted in Figure 15 for the heat treated material containing more than ca. 6 volume percent martensite results from saturation of the ferrite grain interiors with transformation induced dislocation structure. Unless the ferrite grain interiors are maintained relatively dislocation free, the flow stress required to cause the dislocations generated by the austenite to martensite transformation to slip increases and, as measured macroscopically, the yield strength increases. At martensite volume fractions greater than 6%, the yield strength of the lower carbon material increases less rapidly than that of the higher carbon material, with the difference probably due to the larger volume expansion accompanying the austenite to martensite transformation in the higher carbon material.

Continuous yielding can similarly be explained by the effects of the martensite particles on slip initiation at low levels of stress. Comparison of Lüders strain as a function of annealing temperature (Table II) with martensite volume fraction as a function of annealing temperature (Table III) shows that Lüders strain decreases with increasing volume fraction martensite and is zero when the martensite volume fraction exceeds ca. 4%. Yielding occurs uniformly throughout the materials annealed at the higher temperatures because the martensite particles which serve as slip initiation sites (due to transformation induced residual stresses and mobile dislocations) are distributed uniformly throughout the material. Another factor which serves to promote continuous yielding is the inhomogeneous strain distribution shown by Karlsson and Sundstrom (45) to exist in a two-phase microstructure consisting of a continuous ferrite matrix with individual martensite particles. These locally high levels of strain shown by their finite element, continuum mechanics investigation result from a similarly inhomogeneous stress distribution and, since the martensite particles are dispersed uniformly throughout the material, can assist in promoting continuous yielding through each highly stressed microregion acting as a potential slip nucleation site. Lowering of the carbon and nitrogen levels in the ferrite, through partitioning of these elements into the austenite regions during annealing, would further promote continuous yielding by reducing the tendency for mobile, transformation induced dislocations to become pinned by solute atmospheres.

Work Hardening: The rapid increase in flow stress necessary in "dual-phase" steels for meeting GM980X mechanical property specifications requires a high rate of work hardening in the early stages of deformation, i.e., the flow stress of 310-379 MPa (45-55 ksi) at 0.2% strain must increase to 517-552 MPa (75-80 ksi) at 2-4% strain. This investigation shows that the work hardening mechanism in "dual-phase" steels changes with increasing strain in a manner similar to that proposed by Dabkowski and Speich for control-rolled line-pipe steels (9). In the very early stages of deformation, up to ca. 1% plastic strain, very rapid work hardening arises from the same mechanism responsible for low yield strength and continuous yielding, i.e., the transformation induced residual stresses (9, 22, 44) and mobile dislocations. While the residual stresses are being eliminated and the "built-in" mobile dislocations are being immobilized by dislocation-dislocation interactions or by gliding to an interface, the stress required for continued plastic deformation must increase rapidly to a level adequate for the nucleation and propagation of new, mobile dislocations. Therefore, rapid work hardening in this low strain region is due to the continuous transition from a plastic deformation mechanism involving transformation induced residual stresses and mobile dislocations to plastic deformation by a dislocation generation/dislocation glide mechanism.

The next deformation region having a well-defined work hardening mechanism extends approximately from 1% to 7% plastic strain. As shown in Figure 4 for the 1%-5% strain region, the work hardening parameter of the heat treated material varies with annealing temperature and, after annealing at 843°C (1550°F), is a maximum of twice as large as the value for the as-hot-rolled material. The correlation in Figure 16 between the work hardening parameter and the total retained austenite plus martensite "second-phase" content indicates that martensite does not play an important role for work hardening in this region of strain. This conclusion is based on the fact that the materials which were annealed at the highest temperatures and have the largest amounts of martensite (points 4 and 5 in Figure 16, for martensite contents see Table III) show lower work hardening values than do materials which were annealed at lower temperatures and have smaller amounts of martensite (points 2 and 3). In comparison, correlation between work hardening parameter and retained austenite volume fraction (Figure 17) and between work hardening parameter and retained austenite interparticle spacing (Figure 18) shows that retained austenite is the key "second-phase" microconstituent for producing high rates of work hardening in this region of strain.

Explanation of the correlation between retained austenite and work hardening is based on the continuous transformation of the retained austenite to martensite during deformation, indicated by Figure 14, and on the dislocation structure generated by this transformation (7, 8, 20-22). As discussed in detail by Bergstrom (46) and more recently for low carbon steels by Vetter et al. (47), the flow stress in steel is linearly proportional to the square root of the dislocation density. Therefore, since in these "dual-phase" steels a mechanism exists by which the rate of increase in dislocation density per unit strain increase exceeds that for a plain ferrite or a ferrite/martensite steel at the same strain, the rate of work hardening ($d\sigma/d\varepsilon$) for these materials will exceed that for a plain carbon or a ferrite/martensite steel. This situation is demonstrated schematically in Figure 20, where A corresponds to a material with a second phase of only martensite and B corresponds to a material with a second phase which is a mixture of martensite and retained austenite (which is metastable with respect to tensile strain). At least until the retained austenite is nearly totally transformed to martensite, $(d\sigma/d\varepsilon)_B$ will be greater than $(d\sigma/d\varepsilon)_A$ at any given level of strain. Following transformation of nearly all of the retained austenite to martensite, it is expected that the work hardening rate of material B will approach that found for material A at the state of stress found in material B.

This mechanism of increasing the rate of work hardening through a phase transformation process is similar to that encountered in Hadfield steels (48) and metastable austenitic stainless steels (49) and has been recently discussed for high strength austenitic stainless TRIP steels by Pickering (50). Considerable differences do exist between the mechanism by which "dual-phase" steels are strengthened by a martensitic phase transformation and the mechanism by which metastable austenitic steels are strengthened. All of the metastable steels (48-51) are initially 100% austenitic and partially transform upon straining to either a combination of α_{BCC} and ε_{HCP} martensites (for the

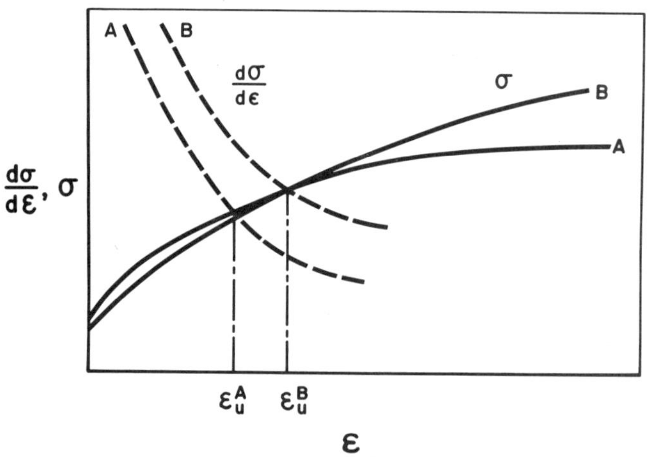

Figure 20. Schematic work hardening ($d\sigma/d\varepsilon$) versus true strain and true stress versus true strain plots for a ferrite + martensite steel (A) and a ferrite + martensite + retained austenite (B) steel.

metastable stainless steels (51)) or to predominately ε_{HCP} martensite (for the Hadfield steel). Strengthening occurs in these steels by replacement of portions of the austenite matrix with stronger martensite and by the grain refinement effect resulting from partitioning of the austenite grains by the martensite plates. As a result of the relatively small volume fraction (< 10 volume %) and isolated particle morphology of the retained austenite in "dual-phase" steels, no grain refinement can occur from austenite transforming to martensite and transformation induced strengthening in the 1% to 6% strain region is due mainly to strengthening of the ferrite matrix by dislocation generation rather than its replacement by martensite.

In the region of strain extending from ca. 8%* to failure, the importance of retained austenite on the rate of work hardening is greatly diminished since the majority of the retained austenite has already transformed to martensite and since the quantity of austenite transforming to martensite per unit strain is much decreased. In this strain region, the "dual-phase" material is essentially a particulate composite (52) consisting of a ductile ferrite matrix containing "hard" martensite and retained austenite particles. The work hardening mechanism in this last region of strain is also that appropriate for a particulate composite material. Although the martensite and retained austenite particles are too large and too widely spaced to produce significant work hardening by the classical mechanism common to precipitation and dispersion strengthened systems (52), recent investigations by Karlsson and co-workers (26, 53) and Araki, Takada, and Nakaoka (54) into the deformation processes of two-phase microstructures (most of the studies concerned the effect of martensite and other types of second phases in steels) have shown that even relatively large and widely dispersed second phase particles can increase the rate of work hardening. Karlsson et al. found that the rate of work hardening increased with increasing volume fraction of second phase (this effect was also found by Matsuoka and Yamamori (6) in ferrite/martensite mixtures) and with increasing difference in hardness between the two phases. Araki, Takada, and Nakaoka (54) recently showed that work hardening in a ferrite/martensite microstructure can be increased by increasing the strength of the martensite.

The mechanism by which a relatively widely spaced second phase can influence the rate of work hardening in a two-phase microstructure is probably related to nonhomogeneous stress-strain distributions developed in such microstructures during deformation. Ashby (55) has shown that compatible deformation of a soft material containing "hard", less deformable particles requires the generation of plastic strain gradients (i.e., gradients of "geometrically-necessary" dislocations) within the more deformable phase. These plastic strain gradients can contribute to work hardening by impeding the motion of dislocations and can thus explain, at least partially, the increased work hardening rate in "dual-phase" steels. The results of Araki et al. (54), which show increasing work hardening with increasing martensite strength, further support this work hardening mechanism since "harder" martensite particles would allow development of greater nonhomogeneous internal stress fields which could more effectively restrict dislocation motion and produce a higher rate of work hardening (9). Assuming good coherency across the ferrite/martensite interface (a necessary condition for compatability during deformation in the finite element continuum mechanics analysis by Karlsson et al. (26)), the transfer, at large plastic strains, of a portion of the load from the matrix to the harder (i.e., stronger) second phase could also contribute to increasing the work hardening rate (53) in "dual-phase" steels.

---

* The exact positioning of the lower boundary of this strain region is rather imprecise and is dependent on both the quantity of retained austenite in a "dual-phase" material and its stability with tensile strain.

Elongation: Steels which undergo a martensitic phase transformation during deformation often exhibit increased rates of work hardening (48-50) and considerably improved elongations (49, 50). As demonstrated by Bressanelli et al. (49), the tensile ductility of a metastable austenitic stainless steel deformed at a temperature at which some martensite forms, i.e., $M_s<T<M_d$, can be significantly greater than that of the same material deformed at $T>M_d$. The improved elongation is derived from the transformation-induced work hardening which produces an increased resistance to the formation of a necked region. Regions of locally high strain, which are likely sites for the nucleation of a neck, are strengthened by undergoing a martensitic transformation and become resistant to increased deformation (49). In another austenitic stainless steel investigation, Tamura and Maki (56) found that the maximum improvement in uniform elongation occurred when the austenite transformed gradually to martensite during deformation.

It is well known that stable plastic flow will continue until the true stress (σ) exceeds the rate at which the material is able to work harden ($d\sigma/d\varepsilon$) (57). Accordingly, plastic flow satisfies the following relationship:

$$d\sigma/d\varepsilon > \sigma \quad \text{stable plastic flow} \qquad [1]$$

$$d\sigma/d\varepsilon \leq \sigma \quad \text{formation of neck region.} \qquad [2]$$

These conditions are illustrated schematically in Figure 20 where true stress and work hardening rate are plotted as a function of true strain. For each of the materials, A and B, the point of instability where $d\sigma/d\varepsilon = \sigma$ and where necking begins, occurs at the point of intersection of the $d\sigma/d\varepsilon$ versus ε and the σ versus ε curves. The strain at this point is the maximum uniform elongation and is designated by ε_u.

The variation in total and uniform elongation for the two materials investigated is shown in Figure 2 as a function of annealing temperature. Up to 843°C (1550°F), the only significant change is the increased uniform elongation (and correspondingly decreased necking elongation) which occurs for both materials. Beyond 843°C, both the uniform and total elongation increase for the lower carbon material and decrease for the higher carbon material. Although work hardening data was obtained only for the 1% to 5% strain region, explanation of the elongation results is possible through consideration of the microstructures developed during annealing and the effects of these microstructures on work hardening. Annealing at 843°C or lower produced a "second-phase" primarily consisting of retained austenite. Comparison of retained austenite volume fraction with uniform elongation, Figure 19, shows a good correlation between increasing retained austenite and increasing uniform elongation. This result is consistent since transformation of the retained austenite to martensite primarily affects work hardening in the early to medium strain region (Figures 14 and 17). The establishment of high work hardening trends in the early and medium strain regions can increase the uniform elongation strain by delaying the onset of necking. If the retained austenite is too unstable and transforms completely to martensite at low strains, the uniform (and probably total) elongation would be low because the flow stress would quickly exceed the rate of work hardening of the material. Conversely, if the retained austenite were too stable the uniform elongation would be low because necking could begin before initiation of the martensitic transformation could increase the work hardening rate of the material. Therefore, the prime elongation effect of the retained austenite is increased uniform elongation. The increase in total and uniform elongation in the lower carbon material after annealing at temperatures above 843°C occurs due to the increased work hardening resulting from the increased volume fraction of martensite. Since the lower carbon material had both a lower YS and UTS and a higher total "second-phase" volume fraction than the higher carbon material, it is expected and experimentally found that the uniform and total elongation of the lower carbon material will exceed that of the higher carbon material.

Conclusions

Heat treatment of both as-hot-rolled materials produced "dual-phase" microstructures consisting of equiaxed ferrite grains with a "second-phase" of discrete retained austenite and martensite particles. The "second-phase" particles formed from austenite which nucleated, during annealing, at the carbides on the ferrite grain boundaries. The volume fraction ratio of martensite to retained austenite, as well as the total volume fraction of "second-phase", increases with increasing annealing temperature. Except for a high density of dislocations in the ferrite regions adjacent to the martensite particles, the majority of the ferrite has a relatively low dislocation density. The martensite particles typically contain fine internal twins characteristic of a high carbon content (probably greater than ca. 0.4 weight %) and a plate transformation mechanism. The retained austenite particles, which are stabilized by a combination of the effects of their small size and carbon enrichment, have an M_s temperature below -269°C and transform to martensite when strained at room temperature.

The annealing treatments produced a large decrease in the yield strengths of both materials without markedly altering their ultimate tensile strengths. Although the effects varied with temperature and carbon content, some annealing treatments produced improved uniform and total elongations. The work hardening parameters of both as-hot-rolled materials were found to be significantly improved by annealing, with ca. a 100% increase resulting from annealing at 843°C (1550°F).

The low yield strength and continuous yielding characteristics of "dual-phase" steels result from the internal residual stresses and mobile dislocations produced by the volume expansion accompanying the transformation of a portion of the austenite (formed during annealing) to martensite during cooling from the annealing temperature.

Rapid work hardening in the very early stages of deformation (up to ca. 1% plastic strain) is due to the continuous transition from a plastic deformation mechanism involving transformation-induced residual stresses and mobile dislocations to plastic deformation by a dislocation generation/dislocation glide mechanism. In the next deformation region, which extends from ca. 1% to 7% plastic strain, work hardening is primarily due to strengthening of the ferrite matrix by the dislocation structure generated during the continuous transformation of the retained austenite to martensite. In the final region of strain, which extends from ca. 8% to failure, the material essentially behaves as a particulate composite and work hardening is due to the development of nonhomogeneous internal stress fields which effectively restrict dislocation motion. Improved uniform and total elongations result from increased stability in plastic deformation produced by the high rate of work hardening.

Acknowledgments

Appreciation is expressed for the many stimulating discussions held with R. A. Kot, W. R. Cribb, S. P. Clough, J. K. Abraham, A. T. Davenport, and R. P. Krupitzer. Grateful acknowledgment is also given to R. Winsteard for development of the retained austenite metallographic etch, to D. A. Baran for quantitative metallographic analyses and to L. J. Vavruska for x-ray measurements.

References

1. M. S. Rashid, SAE Preprint 760206, February (1976).

2. K. Araki, K. Nakaoka, M. Abe and N. Ohashi, 9th Biennial Conf. of IDDRG Research Group Proceedings, 39 (1976).

3. R. L. Cairns and J. A. Charles, JISI 205, 1044 (1967).

4. R. L. Cairns and J. A. Charles, JISI 205, 1051 (1967).

5. R. A. Grange, in Proceedings of the Second International Conference on the Strength of Metals and Alloys 3, Pacific Grove, California, 861 (1970).

6. T. Matsuoka and K. Yamamori, presented at ASM/AIME Materials Engineering Congress, Chicago (1973), published in Met. Trans. 6A, 1613 (1975).

7. S. Hayami and T. Furukawa, Proceedings of the Symposia Micro Alloying 75 IIA, Washington, D.C., 78 (1975).

8. M. S. Rashid, SAE Preprint 770211, February (1977).

9. D. S. Dabkowski and G. R. Speich, Proceedings of the 19th Mechanical Working and Steel Processing Conference, Pittsburgh, Pennsylvania, January 26-27, 284 (1977).

10. D. J. Baily and R. Stevenson, General Motors Technical Center Report GMR-2373, April (1977).

11. K. Araki, S. Fukunaka and K. Uchida, T. ISIJ 17, 701 (1977).

12. R. G. Davies, Met. Trans. 9A, 41 (1978).

13. G. E. Dieter, Mechanical Metallurgy (Second Edition), p. 340, McGraw-Mill Book Co., New York (1976).

14. M. R. Plichta, H. I. Aaronson and W. F. Lange, Metallography 9, 455 (1976).

15. J. M. Rigsbee and W. F. Lange, unpublished research.

16. C. K. H. Dubose and J. O. Stiegler, Rev. Sci. Instr. 38, 694 (1967).

17. W. R. Cribb, Republic Steel Research Center Technical Memorandum TM-12,027-77-5, (1977).

18. R. T. Dehoff, Techniques of Metals Research Vol. 2, R. F. Bunshah, Ed., pp. 221-256, Interscience Publishers, New York, (1968).

19. H. R. Erard, Technical Paper 5A-TR20-2409, Springfield Armory, Springfield, Massachusetts, (1964).

20. K. R. Kinsman, G. Das and R. F. Hehemann, Acta Met. 25, 359 (1977).

21. P. R. Mould, Metals Engr. Quarterly 15, 22 (1975).

22. G. Tither and M. Levite, J. of Metals 27 (9), 15 (1975).

23. G. Krauss and A. R. Marder, Met. Trans. 2, 2343 (1971).

24. G. Thomas, Met. Trans. 2, 2373 (1971).

25. B. I. Edelson and W. M. Baldwin, Jr., T. ASM 55, 230 (1967).

26. B. Karlsson and B. O. Sundstrom, Mat. Sci. and Engr. 16, 161 (1974).

27. H. I. Aaronson, "The Proeutectoid Ferrite and the Proeutectoid Cementite Reactions", in Decomposition of Austenite by Diffusional Processes, p. 387, Interscience, New York, (1962).

28. M. G. Hall, H. I. Aaronson and K. R. Kinsman, Surface Science 31, 257 (1972).

29. K. C. Russell, M. G. Hall, K. R. Kinsman and H. I. Aaronson, Met. Trans. 5, 1503 (1974).

30. J. M. Rigsbee and H. I. Aaronson, submitted to Acta Met.

31. J. M. Rigsbee and H. I. Aaronson, submitted to Acta Met.

32. A. R. Marder, private communication (1977).

33. J. Cahn, Acta Met. 7, 25 (1959).

34. K. C. Russell, Phase Transformations, H. I. Aaronson, Ed., p. 219, ASM, Metals Park (1970).

35. V. Biss and R. L. Cryderman, Met. Trans. 2, 2267 (1971).

36. K. W. Andrews, JISI 203, 721 (1965).

37. W. C. Leslie and R. L. Miller, T. ASM 57, 972 (1964).

38. R. E. Cech and D. Turnbull, T. AIME 206, 124 (1956).

39. C. L. Magee, Met. Trans. 2, 2419 (1971).

40. C. M. Wayman, Introduction to the Crystallography of Martensitic Transformations, MacMillan, New York (1964).

41. C. S. Roberts, T. AIME 197, 203 (1953).

42. C. L. Magee and R. G. Davies, Acta Met. 20, 1031 (1972).

43. J. M. Moyer and G. S. Ansell, Met. Trans. 6A, 1785 (1975).

44. W. C. Leslie and R. J. Sober, T. ASM 60, 459 (1967).

45. B. Karlsson and B. O. Sundstrom, Mat. Sci. and Eng. 16, 161 (1974).

46. Y. Bergstrom, Mater. Sci. Eng. 5, 193 (1969/70).

47. R. Vetter, A. R. Wachters and P. Jongenburger, Scripta Met. 11, 405 (1977).

48. R. A. Hadfield, Metallurgy and Influence on Modern Progress, Chapman and Hall, London, (1925).

49. J. P. Bressanelli and A. Moskowitz, T. ASM 59, 223 (1966).

50. F. B. Pickering, International Metals Reviews, Review 211, 227 (1976).

51. R. B. Benson, R. K. Dann and L. W. Roberts, T. AIME 242, 2199 (1968).

52. H. T. Corten, Modern Composite Materials, p. 27, Addison-Wesley Publishing Co., London (1967).

53. B. Karlsson and G. Linden, Proceedings of the Fourth International Conference on the Strength of Metals and Alloys 1, p. 181, Nancy, France (1976).

54. K. Araki, Y. Takada and K. Nakaoka, T. ISIJ 17 (12), 710 (1977).

55. M. F. Ashby, Strengthening Methods in Crystals, p. 137, John Wiley and Sons, New York (1971).

56. I. Tamura and T. Maki, Toward Improved Ductility and Toughness, p. 183, Climax Molybdenum Development Company (Japan) (1971).

57. G. E. Dieter, Mechanical Metallurgy (Second Edition), p. 342, McGraw-Hill Book Co., New York (1976).

FACTORS AFFECTING THE DUCTILITY OF

"DUAL-PHASE" ALLOYS

A. R. Marder

Research Department
Bethlehem Steel Corp.
Bethlehem, PA 18015

Abstract

An investigation has been conducted to determine the role of the various microconstituents on the ductility of a "dual phase" C-Mn-Si-V steel. It was shown that for the required levels, 50 ksi yield and 90 ksi tensile strength, the optimum ductility was obtained when the volume of the ferrite matrix exceeded 80% and was strain free. In addition, retained austenite was found to be a substantial part of the second-phase constituents and was present in amounts up to 5 volume percent of the structure, or approximately 30-40% of the secondary phases. Although stable down to liquid helium temperature, the austenite was shown to transform on tensile testing. This mechanism, similar to that in a TRIP steel, was confirmed by stepped strain measurements, which showed that the greatest portion of the austenite transformed within the first 10% of strain.

Introduction

Dual phase alloys have recently become of practical interest because they exhibit superior formability to standard high-strength low-alloy steels for a given strength level[1,2]. Most investigations have been concerned with the development of the desired properties, i.e., low yield strength, high tensile strength, excellent work hardening, and ductilities in excess of 25%[3,4]. Recently Davies[5] studied the deformation behavior of these steels and concluded that the superior elongation is largely due to the fine grained high-strength highly ductile ferrite matrix. The present paper reports on the results of an investigation on the microconstituents of a C-Mn-Si-V alloy and their subsequent effect on ductility.

Experimental Procedure

Standard light microscopy and SEM techniques were used to evaluate the microstructure of several alloys (Table 1). A manual point count method with a 100 point grid was utilized in determining the volume percent second phase on light microscopy specimens. Ten areas per specimen at 1000X were measured on as-annealed samples, and several metallographic samples were tempered at 500 F for 30 minutes to better reveal the martensite. Microhardness measurements were made with a Reichart tester and a 7 g load. Standard tensile tests were conducted on 0.100 inch thick samples having a 2 inch gage length using a crosshead speed of 0.1 in./ min up to 4% strain. Thereafter, a crosshead speed of 1.0 in./min was employed until the sample failed. Retained austenite measurements were made by direct comparison of the integrated intensities of austenite and martensite ferrite. Measurements were made for $I_{(200)\alpha}/I_{(200)\gamma}$ and $I_{(200)\alpha}/I_{(220)\gamma}$. Only one heat-treated sample of alloy 2 was available and as a consequence most of the experiments were done on alloys 1 and 3.

TABLE I. ALLOY COMPOSITION

Alloy	C	Mn	P	S	Si	V	Al	N
1	.10	1.32	.010	.007	.54	.11	.097	.0160
2	.12	1.50	.010	.006	.61	.055	.085	.0068
3	.12	1.40	.011	.016	.62	.061	.028	.024

Results and Discussion

The effect of the volume percent ferrite on ductility for all the alloys investigated is shown in Figure 1. As expected, ductility will increase as the percent ferrite increases--or the percent second phase decreases. In all cases there was no pearlite present and all of the steels exhibited continuous yielding. From these results it can be seen that greater than 80% ferrite is needed for superior ductility, i.e., >25% elongation. However, it should be noted that for a given volume % ferrite there exists a large variation in the ductility and even at $\geq$80% volume percent ferrite, ductility can in many cases be less than 25%.

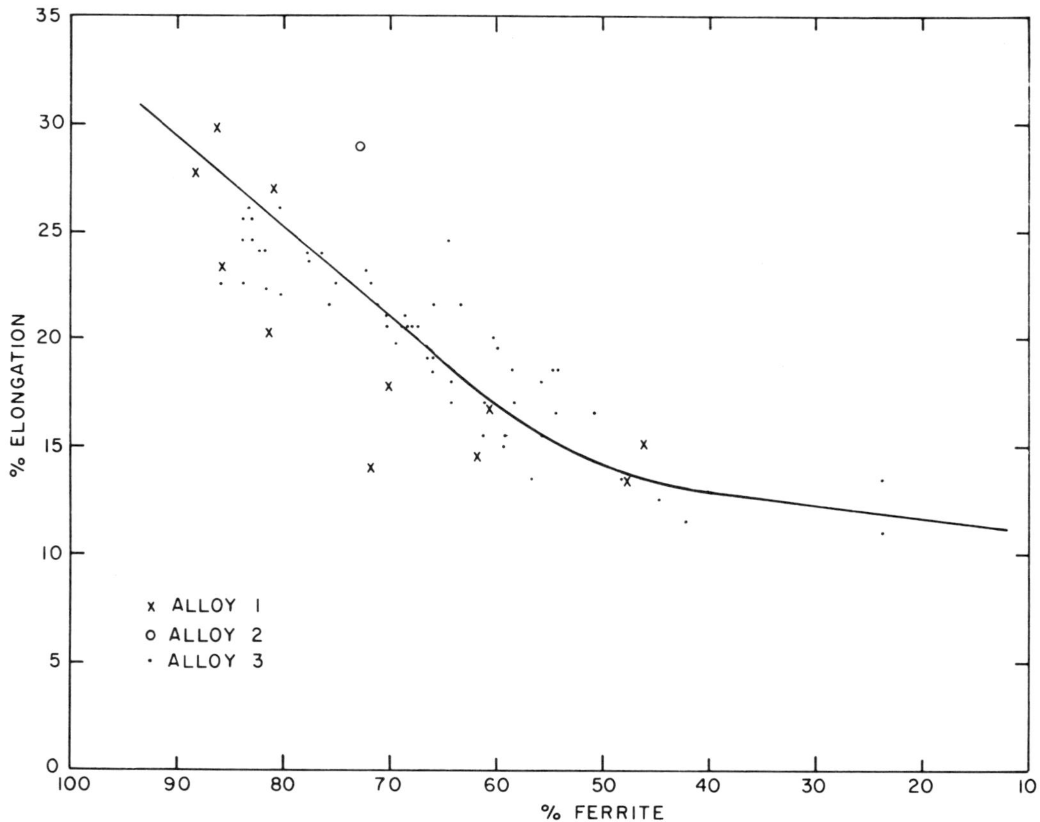

FIGURE 1. The Effect of Volume Percent Ferrite on Elongation

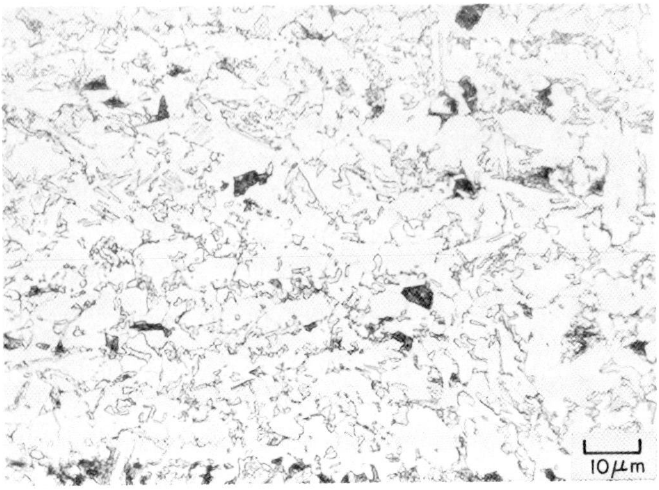

FIGURE 2. Microstructure of Alloy 1 Showing the Variation in Etching Response of Second Phase Constituents. Nital + Sodium Bisulfite Etch.

The deviations in ductility at any fixed volume percent of ferrite may be explained by the microstructure of the "dual phase"alloys, Figure 2. The structure contains a matrix of ferrite together with patches of second phase. The important point to be noted in this photomicrograph is that the second phase does not etch uniformily. Although the etchant used (sodium bisulfite) normally reveals martensite quite well, variations in etching behavior corresponding to variations in martensite with changing carbon content are found. X-ray diffraction analysis showed, however, that up to 5% retained austenite was also present along with martensite in all of the high ductility samples. Thus, the term "dual phase" unless it refers to the heat treatment in the dual phase region, is a misnomer since the alloy contains at least three separate phases: ferrite, martensite and austenite. Both the ferrite and the remaining phases contribute to ductility.

Davies[5] has pointed out that a highly ductile ferrite matrix is essential for superior elongation. Figure 3 shows how increasing the microhardness of the ferrite increases both the yield and tensile strength of Alloy 1, even though the volume percent ferrite is approximately constant. In comparing equivalent volume % ferrite, a 10-20 ksi difference in strength can be found. All the microstructural features, i.e., ferrite grain size, second phase diameter, retained austenite and mean free ferrite path as well as the volume % second phase have been maintained approximately constant. A similar result is found for ductility, Figure 4, where the variation in tensile elongation is approximately 10% for the range of ferrite microhardness studied. The number at each data point represents the retained austenite.

In Figure 4, the low ductility specimens were soaked for 30 seconds at 790 C and water quenched (600 C sec), whereas, the high ductility specimens were soaked for 10 minutes at 790 C and accelerated air cooled (10 C/sec). A comparison of the substructure of these two conditions is seen in Figure 5. In the long soaking time-air cooled-high ductility specimen, Figure 5a, precipitates are found which are too coarse to contribute to the strength

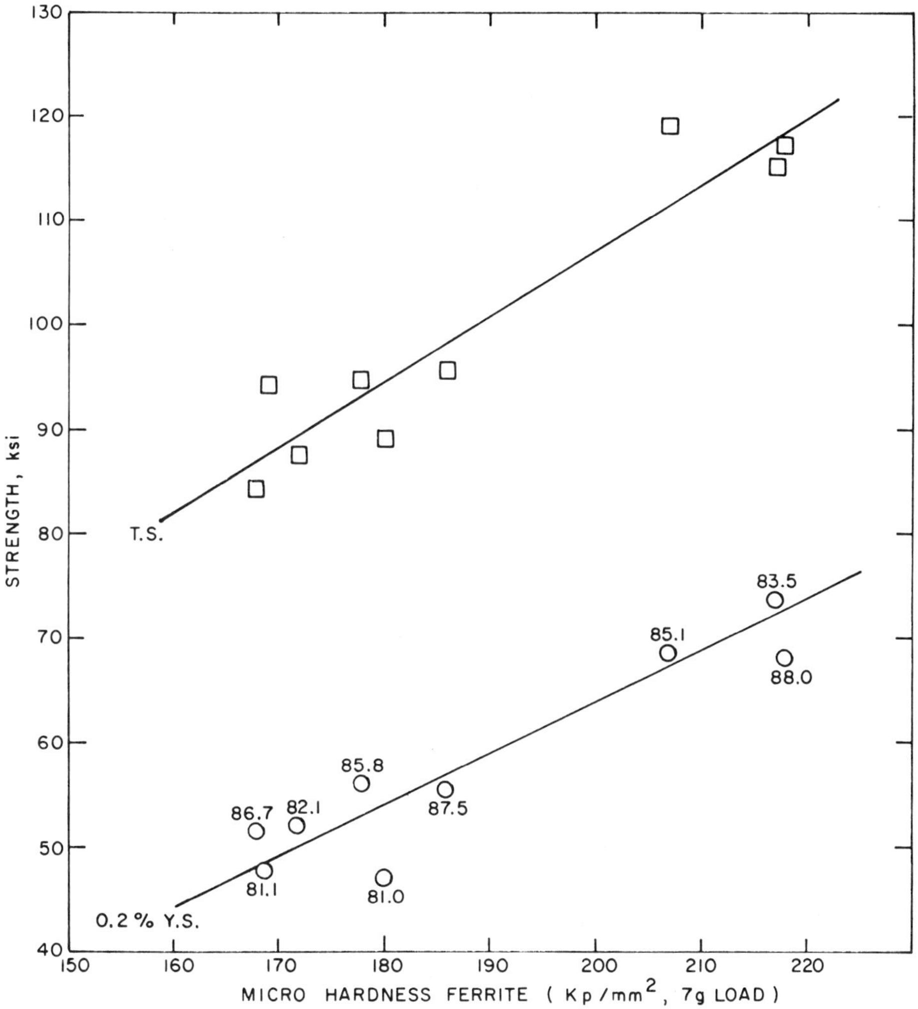

FIGURE 3. The Effect of Ferrite Microhardness on Strength in Alloy 1. Numbers at Each Data Point Represent Volume Percent Ferrite.

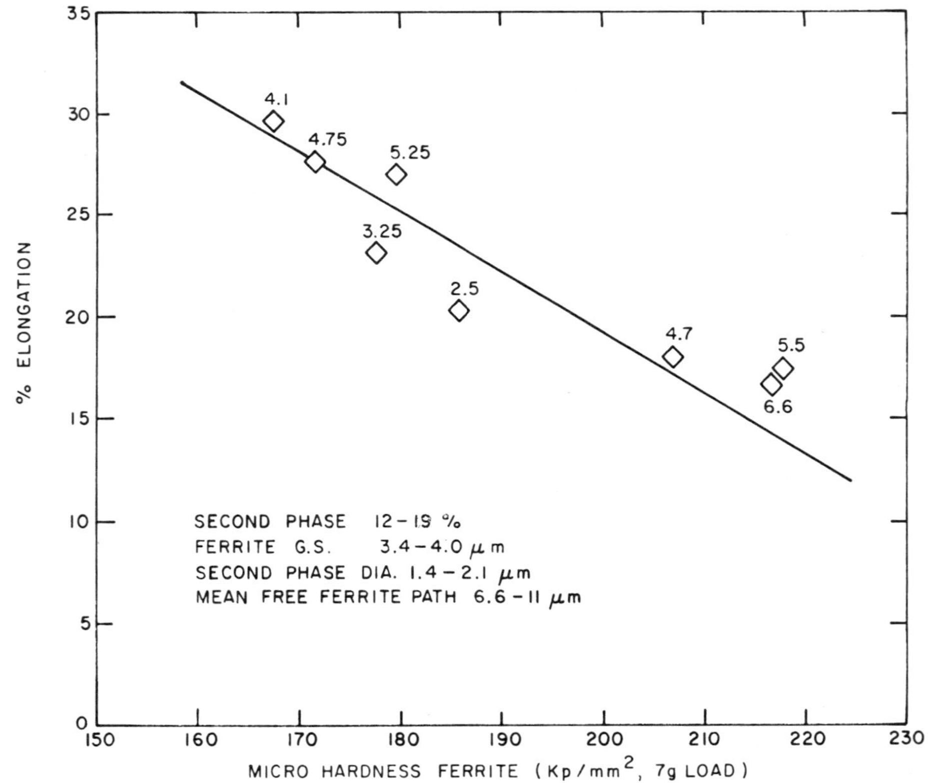

FIGURE 4. The Effect of Ferrite Microhardness on Ductility in Alloy 1. Numbers at Each Data Point Represent Volume Percent Austenite.

of the alloy. For the shorter soaking time - water quench low ductility specimens, Figure 5b, a very fine distribution of precipitates is evident. These precipitates are on the order of 100-200 A and certainly can contribute to strengthening the material and a decrease in ductility. Therefore, the proper intercritical annealing treatment not only must produce a matrix structure that is in excess of 80% ferrite, but the ferrite must have enough time at temperature to enable the fine precipitates produced during hot rolling to coarsen. The large precipitates will then become incapable of contributing to the strength and thus decrease the ductility of the alloy.

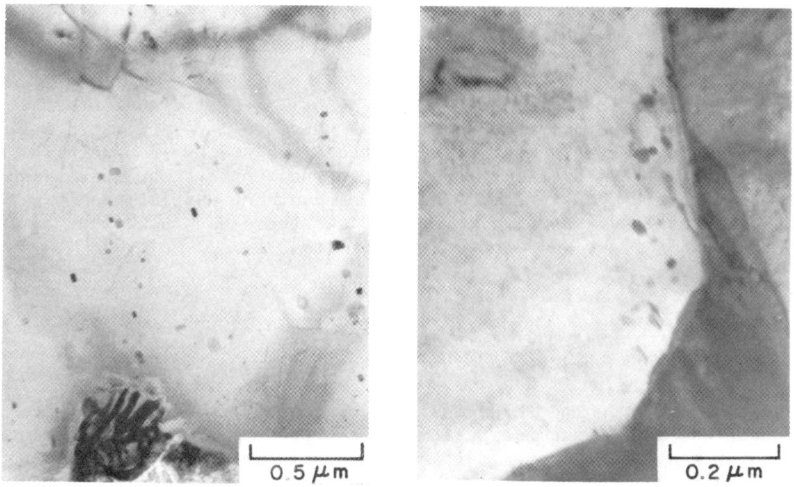

(a) (b)

FIGURE 5. The Substructure of Ferrite in Alloy 1. (a) High Ductility Sample Soaked 10 Minutes at 790 C and Accelerated Air Cooled. (b) Low Ductility Sample Soaked 30 Seconds at 790 C and Water Quenched.

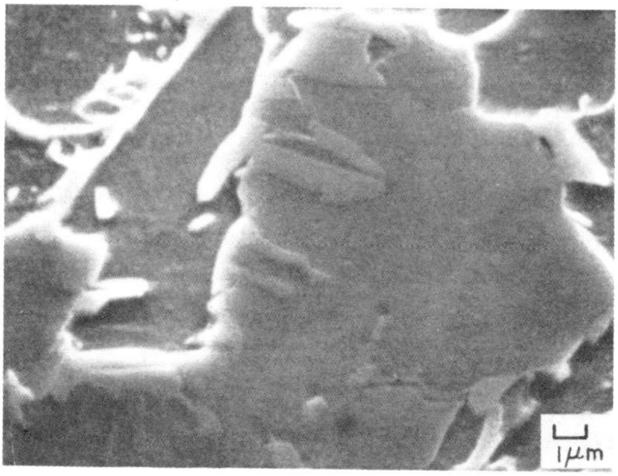

FIGURE 6. Martensite Formation in an Island of Retained Austenite. SEM

In considering the effect of the second phases on ductility, it has already been stated that retained austenite is present, in some cases the second phase contained 30-40% retained austenite. Figure 6 shows the formation of martensite in a patch of austenite. The austenite was found to be extremely stable and did not transform even after quenching in liquid helium, -268.9 C. Only deformation caused the austenite to transform. The variation of retained austenite along a tensile bar of Alloy 2 is seen in Figure 7. The austenite varies from >5% in the grip to <1% at the fractured end. In order to determine the amount of strain necessary to transform austenite, specimens of Alloy 2 were prestrained and the amount of retained austenite measured, Figure 8. It should be noted from this figure that the largest amount of retained austenite transformed within the first 10% of uniform strain. However, even after 10% strain, 2% retained austenite remained in the specimen. The 22% strain represents a total uniform strain in the tensile test.

The transformation of austenite during tensile testing suggests that a "TRIP" mechanism is operative in these "dual phase" alloys and could easily explain the rapid work hardening found in the early stages of testing. This would also suggest that ductility could be further improved by increasing the amount of retained austenite. Figure 9 shows such a relationship between ductility and retained austenite. In all cases there was sufficient time at temperature to effectively soften the ferrite.

Conclusions

The following can be concluded from these experiments:

- In addition to producing in excess of 80% strain free ferrite, the intercritical annealing heat treatment must produce a sufficient amount of retained austenite in the martensite patches in order to obtain the strength-ductility-work hardening relationships necessary for "dual phase" alloys.

- The variation in ductility formed at a constant % second phase is due to the presence or absence of fine precipitates in the ferrite and the amount of retained austenite in the martensite.

Acknowledgements

The author is indebted to A. O. Benscoter and W. J. Noll for their technical support and W. Jolley, B. L. Bramfitt, S. S. Hansen and R. R. Pradhan for their enthusiastic discussions of the subject. The aid of B. S. Mikofsky in preparing the manuscript is also appreciated.

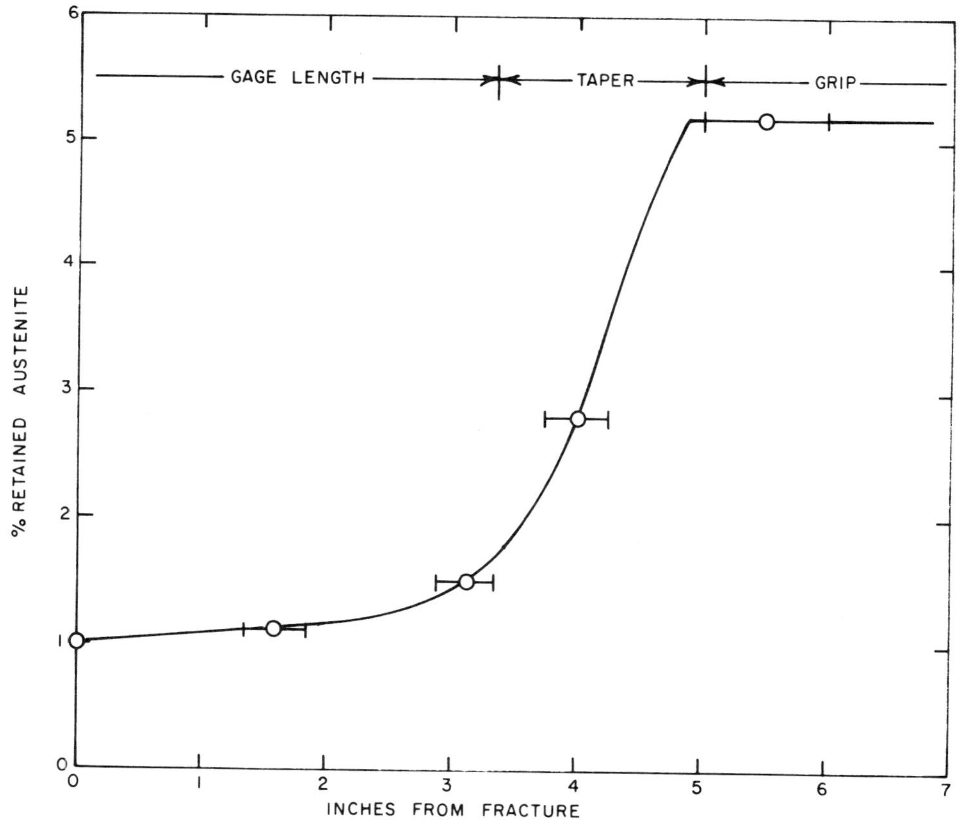

FIGURE 7. The Change in % Retained Austenite in a Tensile Specimen of Alloy 2 After Testing. Error Bar Represents the Size of the Specimen

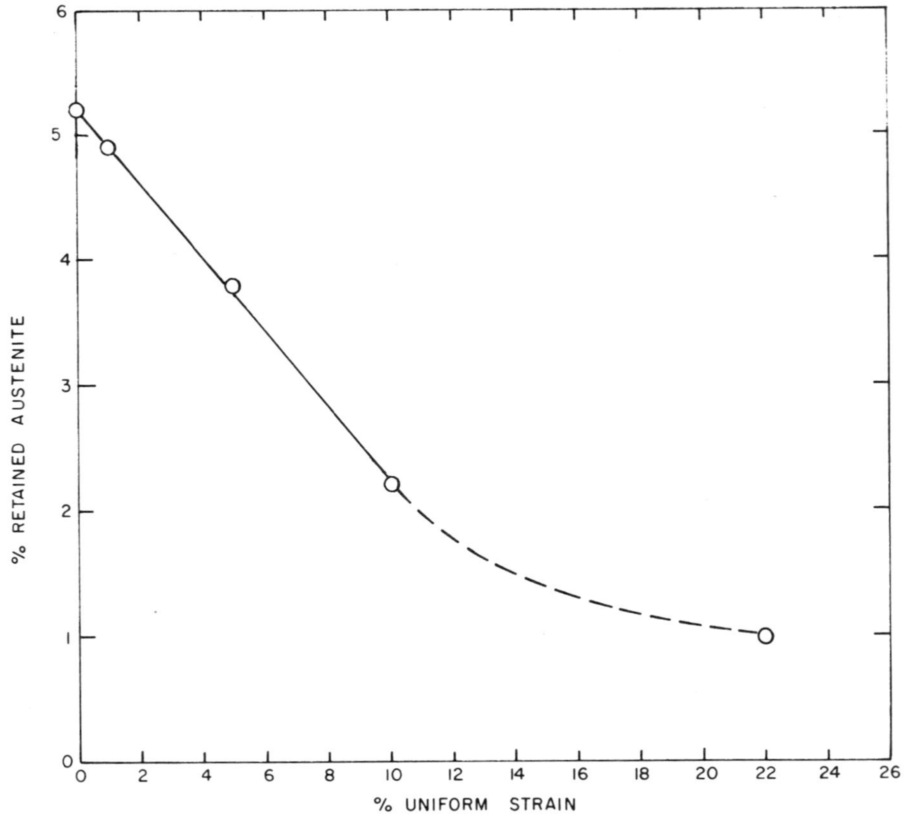

FIGURE 8. The Effect of Prior Strain on Retained Austenite in Alloy 2

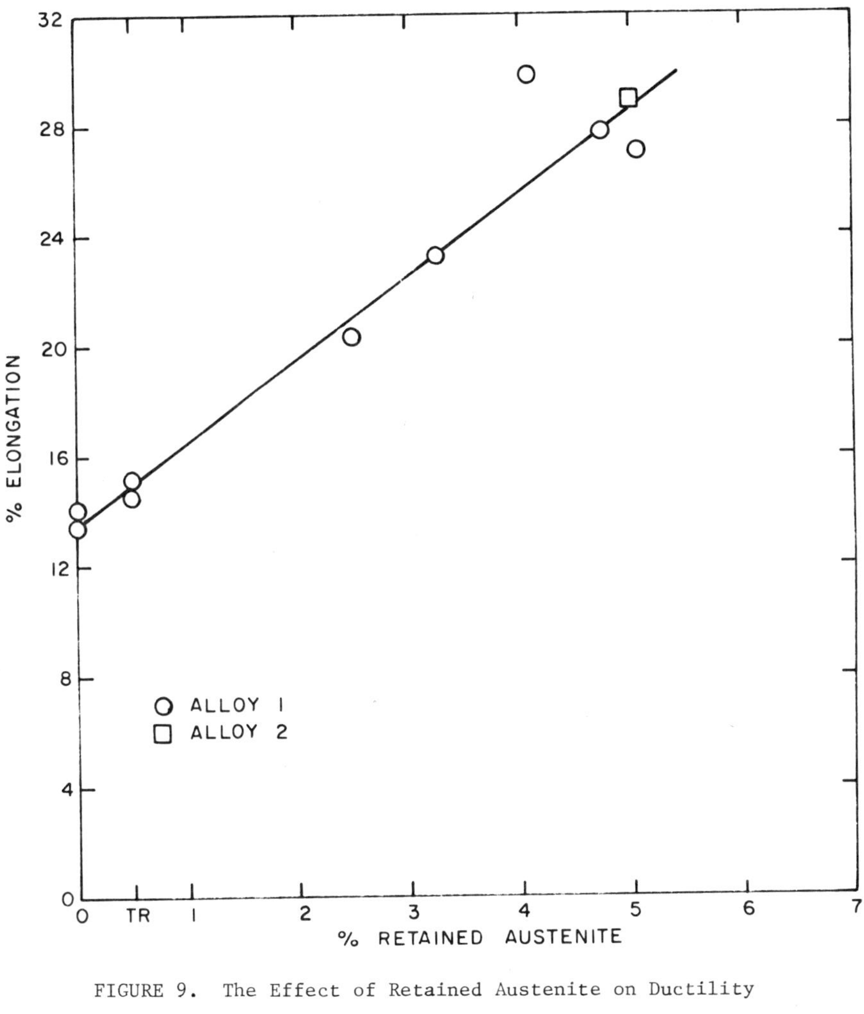

FIGURE 9. The Effect of Retained Austenite on Ductility

REFERENCES

1. M. S. Rashid, SAE Preprint 760206, February, 1976.
2. S. Hayami and T. Furakawa, Proc. Microalloying Conf., Washington, DC, 1975, pp 78-87.
3. J. H. Bucher and G. G. Hamburg, SAE Preprint 770164, February 1977.
4. M. S. Rashid, SAE Preprint 760206, February, 1977.
5. R. G. Davies, The Deformation Behavior of a Dual Phase Alloy, submitted to Met Trans.
6. B. D. Cullity, Elements of X-Ray Diffraction, Addison-Wesley, Reading, Mass, 1956, p 391
7. G. Burwith, Materialpruf., Vol 3, 1961, p 228.

CRACK INITIATION AND PROPAGATION

IN THERMAL MECHANICALLY TREATED SHEET STEELS

Robin Stevenson
Physics Department
General Motors Research Laboratories
Warren, MI 48090

Thermal mechanically treated (TMT) steels have recently been suggested as an alternative to commercial HSLA steels with yield strengths in the range 400-650 MPa. TMT steels derive their strength from a duplex ferrite/tempered-martensite microstructure. This microstructure influences work hardening and significantly modifies the processes of crack initiation and propagation as compared to these processes in more conventional microstructures. In quenched TMT steel cracks initiate by internal fracture of the martensite, while in the quenched and tempered condition the mechanism of crack initiation is decohesion at the ferrite/tempered-martensite interface. In both conditions the crack, once nucleated, propagates through the ferrite. The condition of the material, however, markedly influences the sharpness of the crack tip: extremely sharp cracks are observed in the quenched and tempered condition. The correlation between crack tip sharpness and the macroscopic properties of the ferrite matrix is discussed.

Introduction

Numerous studies of the influence of small volume fractions of second phase on the fracture behavior of a ductile matrix can be found in the literature (1). In addition, some investigators have attempted to separate the crack initiation (2-8) and crack propagation or void growth (7,8) phases of fracture. In general, these studies found that the effect of the second phase was to provide sites for crack initiation, and that the effect of the second phase on crack propagation was minimal for a dispersed noncontinuous second phase. Two mechanisms for crack initiation by a second phase particle have been identified: internal fracture of the particle (2-7) and decohesion between the matrix and the particle (4). In general, a specific second phase particle in a specific matrix will initiate fracture by only one of the above mechanisms. However cases in which both fracture initiation modes occur simultaneously can be found in the literature.

The nature of crack initiation in low carbon steel has been thoroughly investigated. It has been demonstrated that the crack initiation sites are the iron carbide particles and that for elongated carbide particles, the fracture initiation mode is internal fracture of the carbides (2-6). Internal fracture is also observed in spheroidal carbides, but some decohesion has also been observed (4).

The microstructure of TMT steel consists of tempered martensite in a ferrite matrix. Residual carbides are only rarely observed. In view of the novel microstructure of TMT and the influence of crack initiation and propagation on formability, this study was undertaken to determine the crack initiation and propagation behavior of TMT steel. In addition, an attempt was made to separate the contributions of the martensite and the ferrite to the fracture behavior.

Procedure

It has been reported previously that TMT processing can be applied to SAE 1010 (TMT-400), SAE 1010 nitrogenized (TMT-550), and SAE 945X (TMT-650) steels (9). The material employed in this study was a 1.9 mm thick, hot rolled aluminum killed (AK) SAE 1010 steel whose analysis is given in Table I.

Table I. Chemical Composition of SAE 1010 Steel (wt%)

C	Mn	S	Al	O (ppm)	N (ppm)
0.05	0.32	0.016	0.05	30	80

This material was selected to facilitate comparison with the results of previous workers who have examined the fracture of low carbon (ferrite-iron carbide) steels. The similarity between the properties of the family of TMT steels has been noted, and it is anticipated that the results of this study on TMT-400 would also apply to TMT-550 and TMT-650.

The TMT process is indicated schematically in Fig. 1. Samples, precut to ASTM E-8 configuration using a 5.08 mm gage length, were conditioned by immersion in molten salt for a total time of two minutes. (Previous results indicate that this corresponds to a time at temperature of approximately 60 seconds.) The samples were then quenched into an unagitated solution of 5% sodium hydroxide.

To determine the effect of TMT treatment on the ferrite, some samples were "conditioned" at a temperature less than the eutectoid temperature by

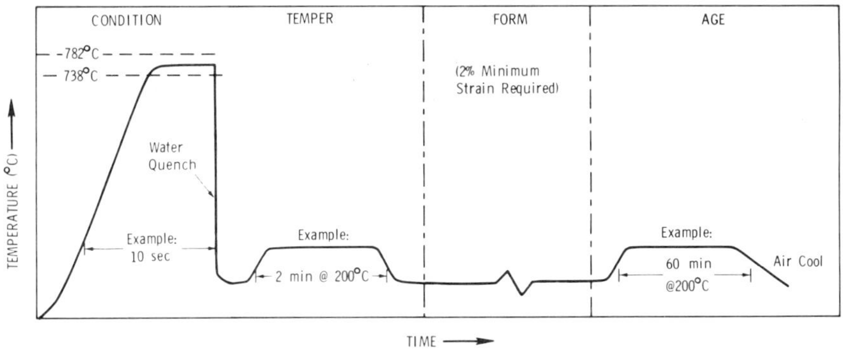

Fig. 1. Schematic of TMT process.

immersion for two minutes in molten salt at 705°C. These samples were quenched as described previously to produce a microstructure consisting of iron carbide in a carbon supersaturated ferrite matrix.

After quenching, some samples conditioned at both 760°C and 705°C were tempered for 10 minutes at 200°C in a temperature controlled oil bath.

All tensile tests were conducted using an Instron tensile test machine. Most specimens were pulled to failure at a constant nominal strain rate of 0.0034/sec, however some samples were used to measure the strain-rate sensitivity using the strain-rate change technique. For these samples alternate strain rates of 0.0017/sec and 0.017/sec were employed. All fractured specimens were examined at a magnification of 1000X in the width and thickness directions using a Reichert MeF1 metallograph.

It was noted previously that all specimens were machined into tensile specimens of ASTM E-8 configuration prior to the conditioning treatment. In addition, some specimens were further machined to produce two notches equidistant from the center of the gage length as shown in Fig. 2. As before,

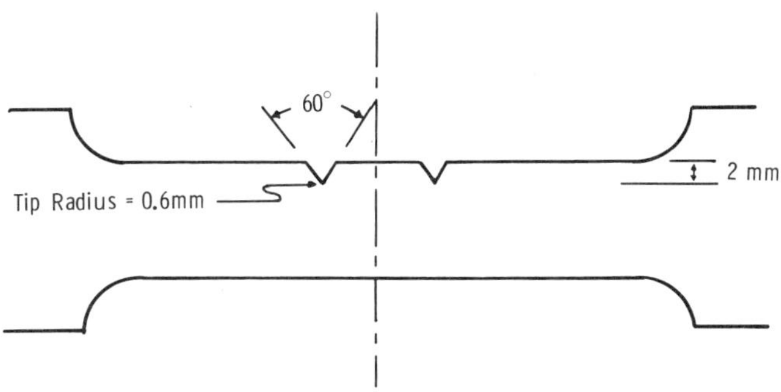

Fig 2. Notched tensile bar.

these specimens were pulled to failure, which invariably occurred at only one of the notches. In these specimens only the area in the root of the unbroken notch was examined metallographically.

Results

The tensile properties and the processing schedules used for each of the materials is summarized in Table II.

Table II. Mechanical Properties of TMT Treated SAE 1010 Steel

	Yield Strength MPa	Tensile Strength MPa	% Uniform Elongation	% Total Elongation (in 5.08 mm)
Conditioned (760°C)	253	457	15.6	20.0
Conditioned (760°C) and Tempered	290	434	18.6	28.8
Conditioned (705°C)	251	392	21.4	29.5
Conditioned (705°C) and Tempered	275	355	19.0	31.8

Metallographic examination of the TMT-400 sample in the as quenched condition indicated that fracture initiated by internal fracture of the martensite colonies and subsequently propagated as a very sharp crack through the ferrite. This is shown in Fig. 3. The effect of tempering on crack initiation and propagation is shown in Fig. 4. Note that two differences are observable between the as-quenched and tempered TMT-400: after tempering, cracks initiate by decohesion between the martensite and ferrite and the cracks which propagate in the ferrite are relatively blunt.

A similar effect of dissolved carbon content on the sharpness of the propagating crack is observed in the material "conditioned" below the eutectoid temperature. In these samples crack initiation invariably occurred by internal fracture of the residual carbide particles. However comparison of the cracking behavior of the as quenched sample (Fig. 5), and the quenched

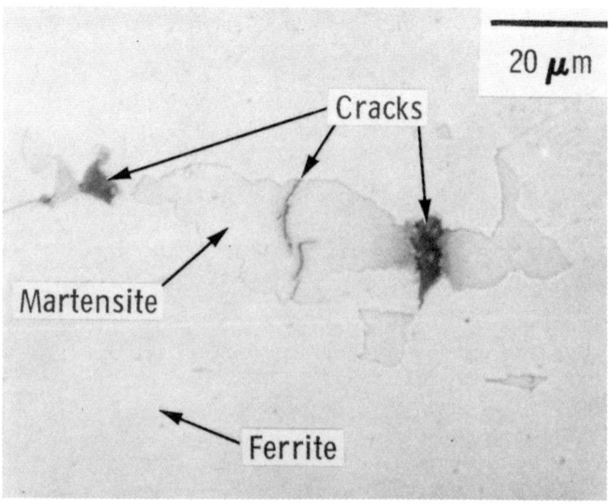

Fig. 3. Fracture observed in sample conditioned at 760°C - as quenched.

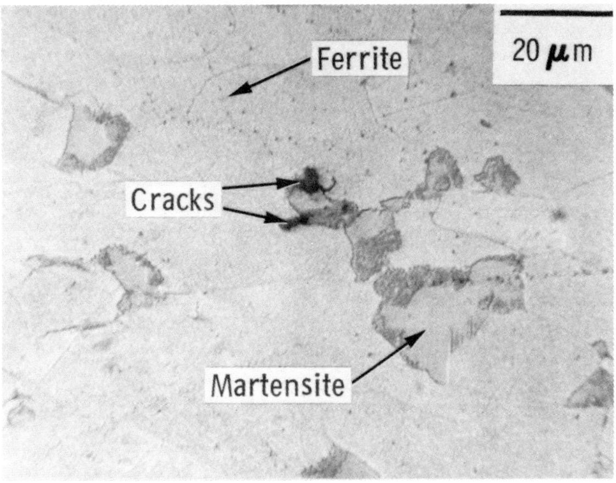

Fig. 4. Fracture observed in sample conditioned at 760°C - tempered.

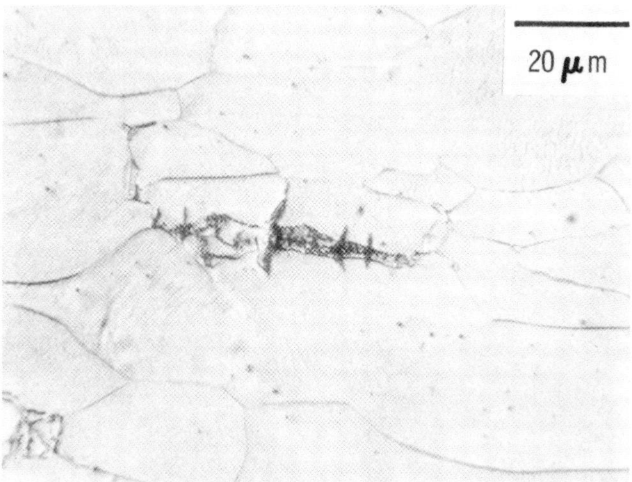

Fig. 5. Fracture observed in sample conditioned at 705°C - as quenched.

and tempered sample, Fig. 6, show the same variation in crack sharpness as observed previously in the TMT-400 samples.

The deleterious effects of the sharp cracks observed in the as quenched samples on fracture toughness are illustrated, in Fig. 7, by the profile of the unfractured notch in double notched TMT-400 tensile specimens.* The differences in notch sharpness between the as quenched and tempered samples

*Recall that fracture occurred at the other notch. Thus the sharpness of the notch profile is a measure of the extent to which flow can be sustained prior to fracture.

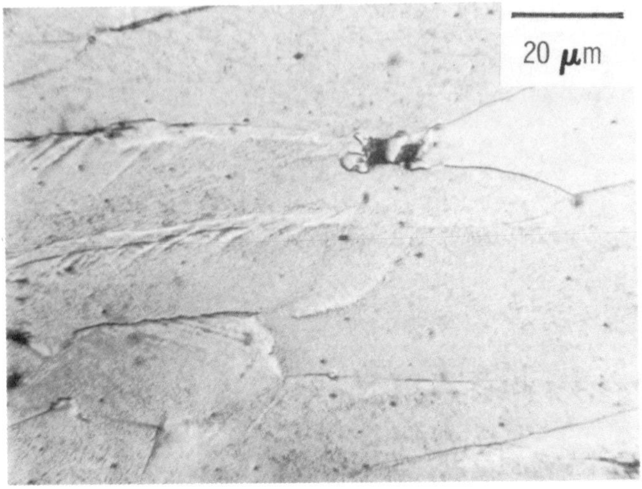

Fig. 6. Fracture observed in sample conditioned at 705°C - tempered.

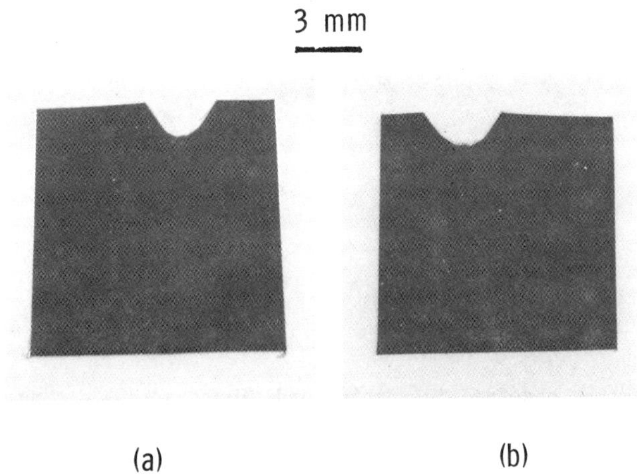

Fig. 7. Notch profiles observed after fracture in samples conditioned at 760°C
a) as quenched
b) tempered.

are obvious. The extent of plastic flow at the root of the notch may be observed more readily in Fig. 8 which shows the distortion of the grains in the as quenched and tempered samples.

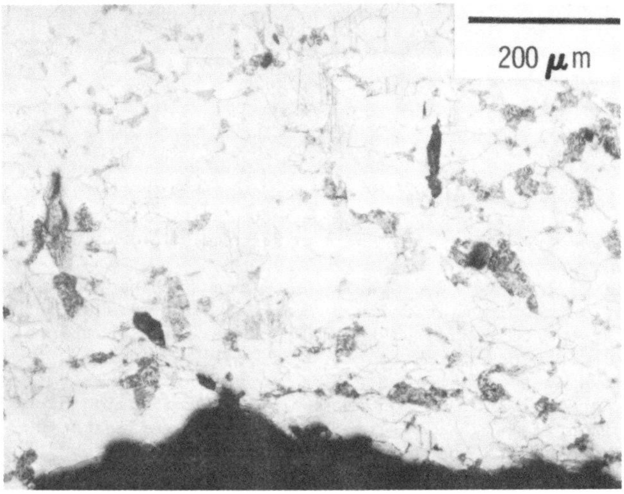

(a)

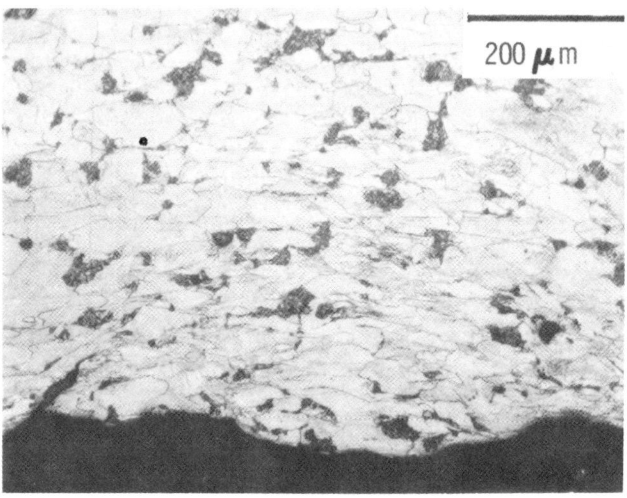

(b)

Fig. 8. Polished and etched section of region at root of notch shown in Fig. 7 - a) as quenched b) tempered.

Figure 9 shows the results of the strain-rate change experiments on samples conditioned below the eutectoid temperature: both as-quenched and tempered. While the tempered specimen exhibits the positive strain-rate sensitivity typically associated with mild steel, the strain rate sensitivity in the as quenched specimen is at best zero and a case can be made for a slightly negative strain-rate sensitivity. In the tempered specimen the measured m value is 0.006-0.008 which is comparable to, but lower than, that generally attributed to mild steel (0.012).

Discussion

Crack Initiation - It is clear that tempering TMT-400 produces a change in crack initiation mode from internal fracture of the martensite islands to decohesion between the martensite and the ferrite. Two mechanisms by which this might occur are:

1) The beneficial effects of tempering on martensite toughness are well documented. Crack initiation is a "weakest link" process, thus simply by improving the toughness of the martensite, some other process (in this case decohesion) which can occur at a lower stress will be favored.
2) The initial stages of martensite tempering involve only the redistribution and segregation of carbon atoms with precipitation occurring on only a limited scale. A preferred site for this segregation will be twin boundaries, lath boundaries, and the martensite-ferrite interface. This carbon segregation would be expected to reduce the interfacial energy at the martensite-ferrite interface and hence to facilitate decohesion.

These mechanisms can thus be distinguished: one increases the stress for martensite fracture relative to the stress for decohesion which is unchanged; and the other lowers the stress for decohesion relative to that for martensite fracture which is unaffected.

However, in the practical situation both mechanisms would be expected to operate simultaneously. Thus, tempering both increases the martensite fracture stress and decreases the martensite/ferrite surface energy. Both effects would lead to the observed change in fracture initiation mode on tempering.

Crack Propagation - It is clear from the parallel studies on the ferrite/martensite structures and the carbide/ferrite structures that the changes in crack propagation behavior are due to a different mechanism than the crack initiation behavior. The changes in crack propagation behavior are related to the dissolved carbon content of the ferrite, as evidenced by the effect of tempering on the samples conditioned at 760°C and 705°C.

As quenched, the dissolved carbon content is high, ~0.02% while previous tempering studies have indicated that tempering reduced the carbon content to ~0.004%. The similarity between the crack propagation behavior in the tempered sample and that observed in annealed hot rolled steel (Fig. 8, Ref. 2) indicates the effectiveness of the lower dissolved carbon content in blunting the cracks.

The shape of the crack on a microscopic scale can be viewed as a measure of the material's ability to distribute strain. A sharp crack is indicative of a material which is unable to distribute strain effectively, while a blunt crack would be observed in a material which can distribute strain effectively. Recognizing the influence of strain-rate sensitivity on strain distribution it is clear that the observed crack morphology correlates well with the measured strain-rate sensitivity.

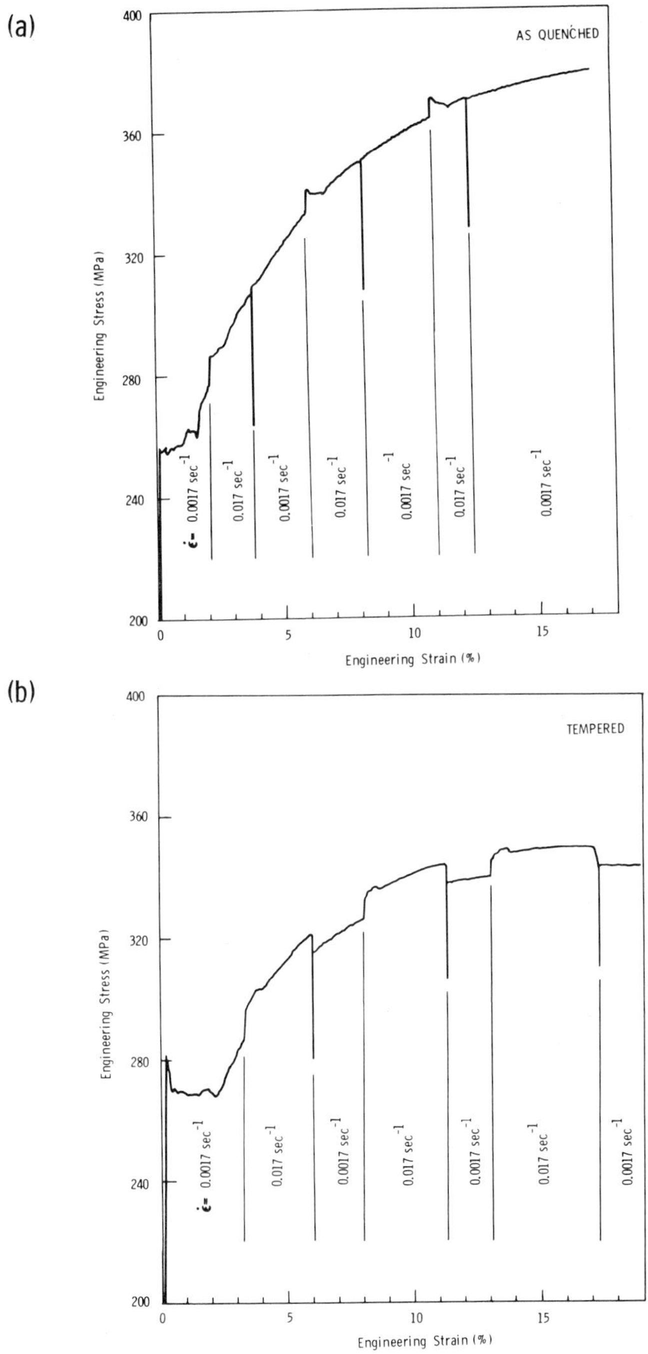

Fig. 9. Chart records illustrating the effect of changing nominal strain rate from 0.0017/sec to 0.017/sec on flow stress in samples conditioned at 705°C - a) as quenched b) tempered.

Thus, the effect of tempering is to reduce the dissolved carbon content which increases the strain rate sensitivity, thereby causing the propagating cracks to blunt. The effectiveness of modifying crack propagation morphology on the macroscopic mechanical properties is not readily measurable. Ghosh (10) has demonstrated the effectiveness of a high strain rate sensitivity in promoting extensive post-uniform elongation and it is appealing to correlate this with the observed changes in crack propagation behavior. However, the strain rate sensitivity also profoundly influences the development of a neck and thus the development of triaxial stresses in the neck which will also influence the propagation of a crack.

However, even if the effect of crack propagation on post-uniform elongation is not readily established it is clear that the effect of tempering on blunting the crack must contribute to the improved elongation of tempered (vs as quenched) TMT-400. This is shown in the notched samples, both by the notch profile, and also by the extent of the flow in the ferrite prior to fracture as shown in the etched samples.

Conclusions

1) Tempering of TMT-400 results in a change in fracture initiation mode from internal fracture of martensite to decohesion at the martensite/ferrite interface.
2) Tempering of TMT-400 causes propagating cracks to blunt due to a reduction in the carbon in solution in the ferrite. This correlates with the effect of tempering on strain-rate sensitivity.

Acknowledgments

The very capable assistance of E. G. Brewer in heat treating and mechanical testing and of T. R. McKinney in the metallography is gratefully acknowledged.

References

1. I. L. Mogford, Metallurgical Reviews, 1967, Vol. 12, p. 49.

2. W. H. Bruckner, Welding Journal, Research Supplement,1950, Vol. 29, p. 467s.

3. W. H. Bruckner, Welding Journal, Research Supplement,1951, Vol. 30, p. 459s.

4. C. T. Liu and J. Gurland, Trans. ASM, 1968, Vol. 61, p. 156.

5. J. Gurland, Acta Met., 1972, Vol. 20, p. 735.

6. C. J. McMahon, and M. Cohen, Acta Met., 1965, Vol. 13, p. 591.

7. J. Gurland and J. Plateau, Trans. ASM, 1963, Vol. 56, 442.

8. H. C. Rogers, Trans. AIME, 1960, Vol. 218, p. 498.

9. D. J. Bailey, SAE Paper No. 760715, presented 18-22 October 1976, Dearborn, MI.

10. A. K. Ghosh, Met. Trans., 1977, Vol. 8A, p. 1221.

EFFECT OF SILICON ON ANNEALING TEXTURE, PLASTIC ANISOTROPY, AND MECHANICAL PROPERTIES OF LOW-CARBON PHOSPHORUS-CONTAINING STEELS

Hsun Hu

U. S. Steel Corporation - Research Laboratory
Monroeville, Pa.

The effects of Si additions to low-C steels, containing 0.04 and 0.07%P, on the annealing texture, plastic anisotropy, and mechanical properties of 80 percent cold-rolled and annealed sheets have been studied. Results show that the yield and tensile strengths of the annealed sheets increased with Si content at a slight sacrifice of the ductility. The excellent r_m and Δr values of the P-containing steels were not adversely affected by Si additions up to 0.70%. With further increase in the Si content (up to 1.24%), the r_m value decreased slightly. Annealing at a higher-than-normal box-annealing temperature, that is, 780°C, resulted in higher r_m values but lower strength. Conversely, annealing at a lower temperature resulted in lower r_m values but higher strength. Thus, by proper selection of the steel composition and annealing temperature, a wide variety of strength and drawability combinations can be produced. The increase in strength is due to solution strengthening and grain refinement by P and Si in the present steels. Little interaction between P and Si is indicated by the results obtained.

Introduction

Previous studies have shown that the addition of phosphorus to low-carbon low-manganese steels improves their deep-drawing properties and strength,[1] when the hot-rolled band is cold-rolled 80 percent and simulated box-annealed at 710 or 780°C (1310 or 1435°F) for 20 hours. These excellent properties of phosphorus steels are not adversely affected by increasing the manganese content up to 0.25 percent,[2] and the carbon content up to 0.05 percent.[3] Further improvement in yield strength (up to 345 to 379 MPa or 50 to 55 ksi) at some sacrifice of the drawability (r_m values of 1.3 to 1.8) and ductility can be achieved through grain refinement[4] by annealing at low temperatures or by utilizing a rapid heating and an abbreviated annealing cycle to simulate the conventional continuous annealing, or by electrical induction annealing with high heating rates.[5] An additional increase in strength can be obtained by subsequent strain-aging, but a substantial reduction in ductility occurs as a consequence. All these results apply to vacuum-melted as well as air-melted steels.

To obtain further information about the characteristics of phosphorus steels, the effects of silicon content on the annealing texture, plastic anisotropy, and mechanical properties of these steels have been studied. Silicon is an effective solution strengthener in iron and steel,[6] and the addition of silicon should further increase the strength of phosphorus steels. Therefore, a study of the effect of silicon on the annealing texture, hence the drawability, of phosphorus steels and the maximum limit of the silicon content for the best compromise in drawability and strength would be of great practical value and of scientific interest.

Materials and Experimental Work

Chemical Compositions

A 136-kg (300 lb) vacuum-melted heat with an aim base composition of 0.04 percent C, 0.20 percent Mn, and 0.07 percent P was made in the laboratory. The heat was split-cast into six ingots having levels of silicon content ranging from 0.02 to 0.70 percent. The base composition was selected, on the basis of previous investigations,[1-3] to produce a combination of excellent deep-drawing and moderately high-strength properties in cold-rolled and box-annealed sheets.

The chemical compositions of the steels investigated are shown in Table I. As can be noted, the actual carbon (about 0.02%) and phosphorus (about 0.04%) contents of the steels were considerably below the aim composition, although the manganese content was almost exactly as specified. The oxygen contents in Steels A and B (the steels with the lowest silicon content) varied considerably between the top and bottom of the ingots.

For most of the other steels, such inhomogeneities were negligible.

Table I

Chemical Composition of the Vacuum-Melted Low-Carbon Phosphorus Steels—Weight Percent

Steel 7364-8029		C	Mn	P	S	Si	Cu	Ni	Cr	Al_{sol}	Al_{total}	N	O_{ppm}
A	T	0.027	0.192	0.037	0.022	0.021	<0.005	0.020	0.018	<0.001	0.002	0.004	219
	B	0.020	0.201	0.043	0.021	0.021	0.005	0.020	0.020	<0.001	<0.003	0.004	93
B	T	0.025	0.196	0.047	0.020	0.032	<0.005	0.020	0.018	0.001	0.005	0.004	127
	B	0.028	0.199	0.048	0.021	0.042	0.005	0.020	0.020	<0.001	<0.002	0.003	92
C	T	0.025	0.195	0.047	0.021	0.079	<0.005	0.022	0.018	0.001	0.003	0.004	70
	B	0.024	0.188	0.046	0.022	0.078	0.005	0.022	0.020	<0.001	<0.002	0.003	78
D	T	0.026	0.200	0.047	0.022	0.194	<0.005	0.020	0.020	<0.001	<0.002	0.003	55
	B	0.014	0.201	0.043	0.019	0.183	0.005	0.022	0.020	<0.001	0.002	0.006	52
E	T	0.023	0.197	0.048	0.023	0.355	<0.005	0.020	0.020	<0.001	<0.002	0.003	50
	B	0.020	0.205	0.044	0.020	0.365	0.007	0.022	0.020	<0.001	0.002	0.004	42
F	T	0.022	0.198	0.047	0.021	0.660	<0.005	0.020	0.020	<0.001	<0.002	0.003	56
	B	0.018	0.203	0.044	0.020	0.700	0.007	0.020	0.020	<0.001	<0.002	0.003	52

T = Top of Ingot. B = Bottom of Ingot.

One additional vacuum heat (114 kg or 250 lb), split-cast into five ingots having various levels of silicon, was made later in the investigation. The purpose of including these additional steels in the investigation was threefold: (1) to extend the silicon content to still higher levels so that a preferred range of silicon for good deep-drawing and high-strength properties could be established; (2) to correct the base composition of the earlier steels (Table I), in particular the carbon and phosphorus contents, to the specified levels; and (3) to affirm the reproducibility of the results obtained from the steels of the first heat.

The chemical compositions of the steels made from the second heat are shown in Table II. As can be seen, the silicon content was extended to a higher maximum level than in the earlier steels

Table II

Chemical Composition of the Vacuum-Melted (Second Heat) Phosphorus Steels—Weight Percent

Steel 7464-8006		C	Mn	P	S	Si	Cu	Ni	Cr	Al_{sol}	Al_{total}	N	O_{ppm}
A	T	0.032	0.204	0.072	0.021	0.087	0.010	0.023	0.024	<0.001	<0.002	0.003	111
	B	0.034	0.192	0.070	0.021	0.091	0.010	0.023	0.024	<0.001	<0.002	0.003	125
B	T	0.034	0.202	0.072	0.022	0.276	0.012	0.023	0.024	<0.001	<0.002	0.003	77
	B	0.032	0.195	0.070	0.021	0.263	0.012	0.022	0.024	0.001	<0.002	0.003	79
C	T	0.033	0.199	0.071	0.022	0.545	0.012	0.023	0.024	0.001	0.004	0.002	68
	B	0.033	0.191	0.070	0.021	0.539	0.012	0.022	0.024	<0.001	<0.002	0.002	71
D	T	0.033	0.202	0.070	0.021	0.880	0.012	0.023	0.024	<0.001	0.004	0.003	53
	B	0.031	0.200	0.069	0.021	0.875	0.012	0.022	0.024	0.001	0.003	0.002	62
E	T	0.031	0.194	0.070	0.021	1.247	0.012	0.022	0.024	<0.001	0.002	0.003	42
	B	0.031	0.194	0.069	0.020	1.231	0.012	0.022	0.024	0.001	0.002	0.003	38

T = Top of Ingot. B = Bottom of Ingot.

(1.2% vs 0.7% Si). The carbon, manganese, and phosphorus contents were all fairly close to the specified levels. The size of the ingots of all these steels was 75 by 140 by 250 mm (3 by 5.5 by 10 in.).

Hot-Processing Procedures

The hot-processing procedures were essentially the same as those described in earlier investigations.[1-3] The ingots were heated to 1230°C (2250°F) and hot-forged on all sides to approximately 70 by 110 mm (2.75 by 4.50 in.) in cross section. The forged pieces were later reheated to 1230°C, hot-rolled from 70 to 12.5 mm (2.75 to 0.50 in.) in thickness, and air-cooled. Samples for analyses were obtained at this stage by cutting out pieces at locations corresponding to the top and bottom of the ingot. Final hot rolling was conducted by reheating the 12.5-mm-thick plates to 1230°C, and hot rolling to 3.8 mm (0.150 in.) with the finishing pass at 925°C (1700°F). The hot-rolled band was water-spray-cooled to approximately 620°C (1150°F), then furnace-cooled at a rate of about 40°C (75°F) per hour to simulate the cooling conditions in commercial coils.

Cold Rolling and Annealing

The hot-rolled bands, 3.8 mm thick, were sandblasted to remove surface scale, and were pickled to clean the surface. They were then cold-rolled to approximately 80 percent reduction in thickness to 0.8 mm (0.030 in.). Tension specimens were machined from blanks cut from the cold-rolled strips at 0, 45, and 90 degrees to the rolling direction. These specimens were then annealed in loose packs in an atmosphere of 15 percent $H_2 + N_2$ by heating at a rate of 20 to 25°C (40 to 50°F) per hour to 650, 710, or 780°C (1200, 1310, or 1435°F), held at temperature for 20 hours, and furnace-cooled (simulated box annealing).

The anisotropy parameters, $r_m = (r_0 + 2r_{45} + r_{90})/4$, and $\Delta r = (r_0 - 2r_{45} + r_{90})/2$, where the subscripts denote the angles from the rolling direction to the strip, along with other mechanical properties of the annealed specimens, were determined from uniaxial tension tests in the usual manner.

Results and Discussion

Microstructure and Texture of Hot-Rolled Bands

The through-thickness microstructure of the 12.5-mm-thick hot-rolled bands was examined metallographically on longitudinal and transverse cross sections. The general features of the microstructure were largely similar in that they all exhibited complete recrystallization and fairly equiaxed grains. In most of the specimens, the grain size was fairly uniform across the thickness of the hot band. For specimens having relatively low silicon contents and annealed at the high temperature (780°C), a few coarse grains were found in the finer grained matrix. The average grain size varied only slightly, ranging from 9 to 10 ASTM, among the steels.

The textures of the hot-rolled bands, as indicated by the X-ray intensities of the various reflections measured from the surface, quarterthickness, and midthickness sections, were not perfectly random. There was a tendency for the (200) intensity to become stronger on moving from the surface to the midthickness of the specimen. The (211) and (222) reflections generally indicated a maximum at the quarterthickness section. On the other hand, the (310) intensity varied little through the thickness of the hot-rolled band. However, overall, the hot-rolling texture was very weak; and based on the light microscopy and texture analysis, silicon content appears to have little effect on the hot-rolling deformation and dynamic recrystallization of the austenite under the hot-processing conditions employed. The texture observed in the hot-rolled band is, of course, the texture of the ferrite, a product of phase transformation from the hot-rolled austenite upon subsequent cooling.

Texture of the As-Cold-Rolled Strips

The crystallographic texture of the cold-rolled strips, after a reduction in thickness of approximately 80 percent from 3.8 to 0.8 mm (0.150 to 0.03 in.), was examined by X-ray intensity measurements from the surface and the midthickness sections. The cold-rolling texture varied little among the steels with various silicon contents, and is characterized by strong (222), (200), and (211) intensities, typical of the cold-rolling texture of iron and low-carbon steels. These strong reflections correspond to the major components of (111)[$\bar{1}\bar{1}2$], (112)[$\bar{1}10$], and (100)[011] type orientations in the rolling texture of bcc metals and alloys. There were only small variations in the intensities of textures at the surface and the midthickness sections.

Texture of the Annealed Strips

The textures of the annealed specimens were examined at the surface and the midthickness of the strip by X-ray intensity measurements of the various reflections. For some of the specimens, X-ray pole figures were also determined from the midthickness section.

Results of the X-ray intensity measurements of the various reflections indicated that, in general, annealing at the higher temperatures resulted in stronger (222) and weaker (200) orientations. The effect of silicon on the annealing texture of phosphorus-containing steels was very small, and a readily detectable difference was observed only between steels differing substantially in silicon content. Typical examples of the pole figures are shown in Figures 1A, 1B, and 1C for a steel containing 0.88 percent silicon (Steel D, Table II), and annealed at 650, 710, and 780°C respectively. The nature of the texture is

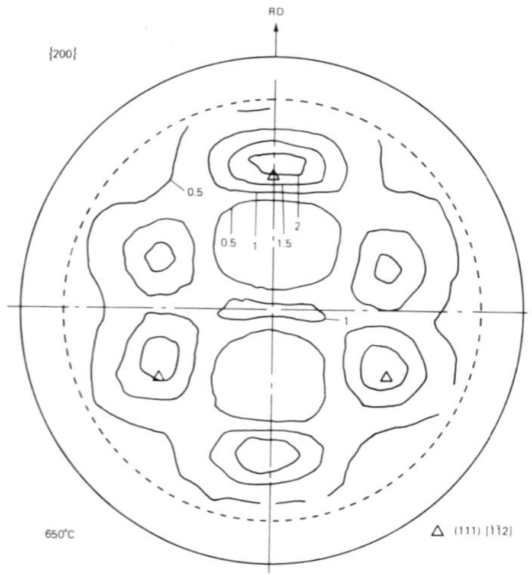

POLE FIGURE OF LOW-CARBON STEEL CONTAINING 0.07%P
AND 0.88%Si ANNEALED AT 650°C (1200°F)

Figure 1A.

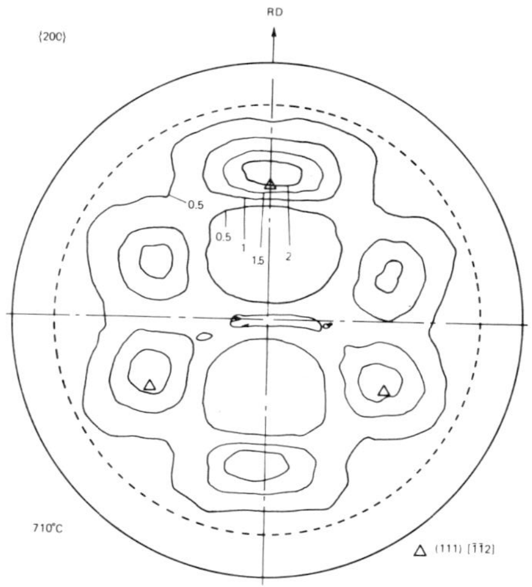

POLE FIGURE OF LOW-CARBON STEEL CONTAINING 0.07%P
AND 0.88%Si ANNEALED AT 710°C (1310°F)

Figure 1B.

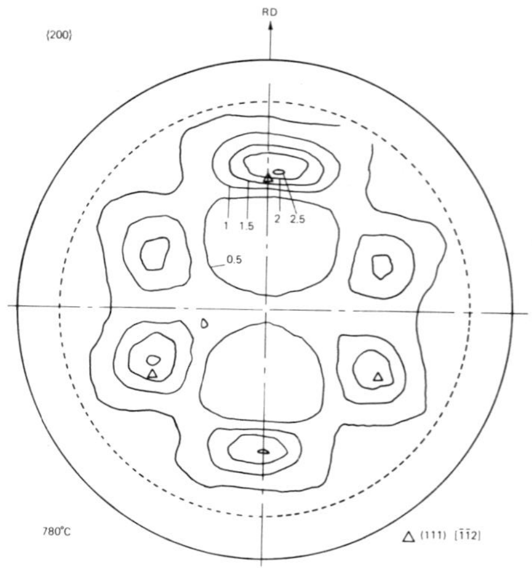

POLE FIGURE OF LOW-CARBON STEEL CONTAINING 0.07%P
AND 0.88%Si ANNEALED AT 780°C (1435°F)

Figure 1C.

the complementary (111)[$\bar{1}\bar{1}2$] and (111)[$11\bar{2}$] orientations, characteristic of the phosphorus-containing steels.[1,2]

The intensity of the texture developed at 780°C (Figure 1C) is somewhat stronger than those developed at the lower temperatures (Figures 1A and 1B). For previously studied[1,2] steels containing a similar amount of phosphorus but with much lower silicon contents (0.01 to 0.03% Si), the intensities of the {111}<112>-type texture were considerably stronger (maximum intensity levels, 3 to 5 times random) than the pole figures shown in Figure 1. These observations indicated that the r_m value, or drawability, of the phosphorus-containing steels tends to decrease at high silicon concentrations (above 0.8 to 0.9% Si) and at low annealing temperatures (<710°C).

The r_m Value and Yield Strength

Since the drawability and strength of the annealed strips are of primary interest in the present investigation, the average plastic-strain ratio, or r_m value, and the yield strength of the annealed strips are presented as a function of silicon content in Figures 2A, 2B, and 2C for specimens annealed 20 hours at 650, 710, and 780°C, respectively. As shown by these figures, the yield strength increased linearly with increasing silicon content. The data points for the steels made from the two separate heats can be represented by a single line, with some scatter only at the low end of silicon contents. This indicates that the results of the present investigation had good reproducibility. The data scatter at low silicon concentrations was likely a consequence of variations in the base composition.

There was a tendency for the r_m values to increase with increasing silicon content at the low end of silicon concentrations, but a gradual decrease of the r_m value with increasing silicon was definitely indicated at high silicon contents (above about 0.6%). For a given silicon content, the r_m value decreased with decreasing annealing temperature. These observations are all consistent with the texture data described in a previous section. For the range of silicon studied, the yield strengths of the annealed strips increased linearly with silicon content at a rate of approximately 100 to 120 MPa (15 to 18 ksi) per 1 percent silicon. At a concentration of 1 percent silicon, the yield strength would increase 50 MPa (7 ksi) from 300 to 350 MPa (44 to 51 ksi), and the r_m value would decrease 0.4 units, from 2.0 to 1.6 with a decrease in the annealing temperature from 780 to 650°C.

It was pointed out previously[1] that the excellent r_m values produced in phosphorus-containing steels was believed to be due to preferential segregation of phosphorus at the subgrain and grain boundaries. The small amounts of residual soluble aluminum in these steels could not be responsible for the deep-drawing texture developed during the simulated box-annealing treatment. Evidence in support of this statement was presented and discussed

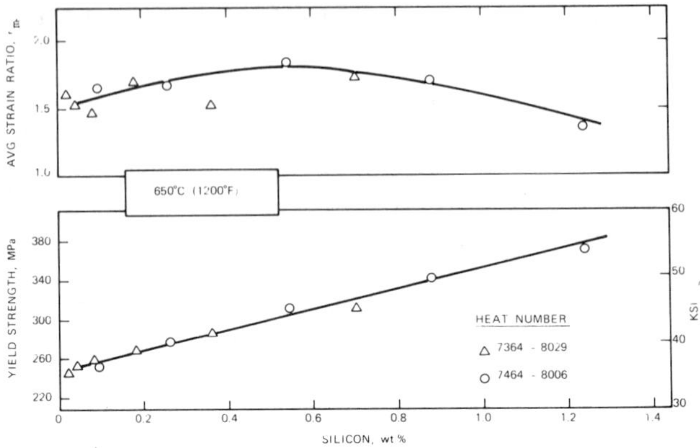

EFFECT OF SILICON ON r_m VALUE AND YIELD STRENGTH OF VACUUM-MELTED LOW-CARBON PHOSPHORUS STEELS ANNEALED AT 650°C

Figure 2A.

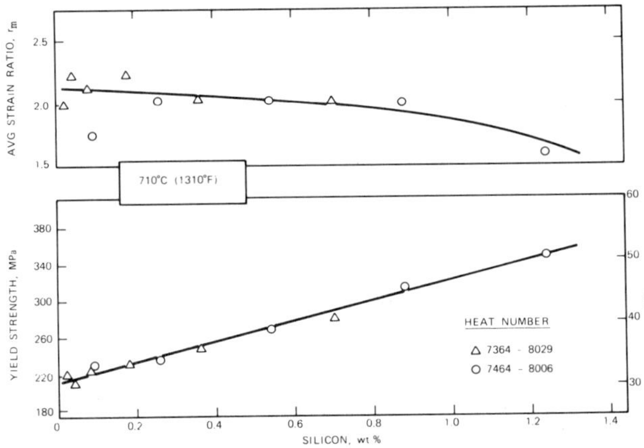

EFFECT OF SILICON ON r_m VALUE AND YIELD STRENGTH OF VACUUM-MELTED LOW-CARBON PHOSPHORUS STEELS ANNEALED AT 710°C

Figure 2B.

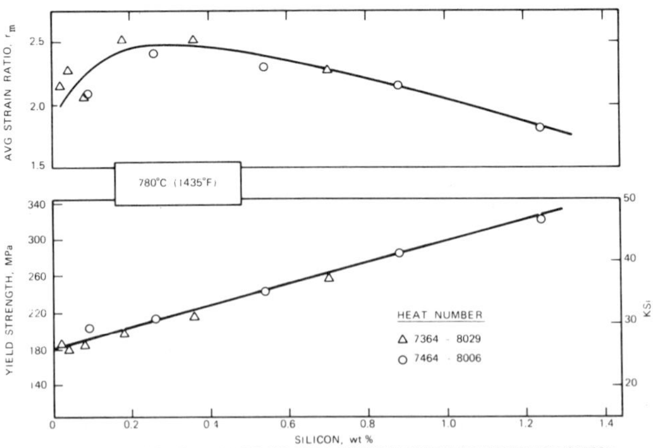

EFFECT OF SILICON ON r_m VALUE AND YIELD STRENGTH OF VACUUM-MELTED LOW-CARBON PHOSPHORUS STEELS ANNEALED AT 780°C

Figure 2C.

on five points, which are equally applicable to the present phosphorus-silicon steels and need not be restated here.

Effect of Degree of Straining on r_m Value

The r_m values shown in Figures 2A, 2B, and 2C were determined from specimens tensile-strained to 15 percent elongation, which corresponded to approximately the midpoint of uniform deformation. Recent studies[7,8] have shown that the r values are strain-dependent, and the variation in r value with increasing strain depends on the texture of the specimen. For (111)-textured sheet, the r value decreases with increasing strain. Figure 3 shows a comparison of the r_m values determined from specimens tested to maximum load (25 to 30% elongation) with those from specimens strained to 15 percent elongation. Appreciably lower r_m values were obtained from specimens tested to higher strains. Thus, these results are in agreement with the predictions of the earlier studies.[7,8]

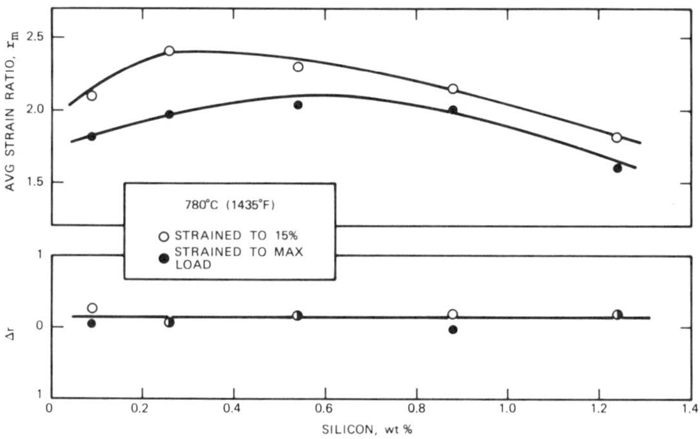

EFFECT OF AMOUNT OF STRAINING ON r_m AND Δr VALUES OF VACUUM-MELTED LOW-CARBON PHOSPHORUS STEELS ANNEALED AT 780°C

Figure 3.

In the same figure, the planar-anisotropy parameter, Δr, was shown to be very little affected by the silicon content of the steel, which is consistent with the texture data in that the nature of the annealing texture was essentially the same for all the steels studied, only the intensity of the texture being different. The same plot also shows that the amount of straining had very little effect on the Δr values, even though the r_m values were influenced markedly. These observations appear to be in support of the interpretation that the strain dependence of the r value is not due to the orientational changes of the specimen during tensile deformation, but arises from anisotropic work hardening in textured sheets.[9] The very low Δr values of these phosphorus-silicon steels suggest weak tendencies for earing in cup-drawing operations.

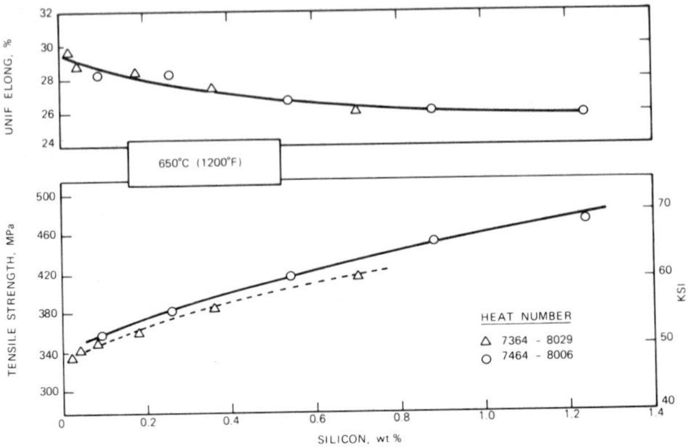

EFFECT OF SILICON ON THE ULTIMATE TENSILE STRENGTH AND UNIFORM ELONGATION OF VACUUM-MELTED LOW-CARBON PHOSPHORUS STEELS ANNEALED AT 650°C

Figure 4A.

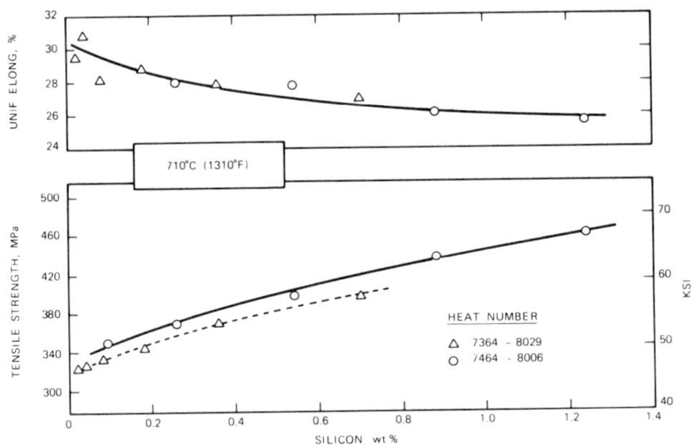

EFFECT OF SILICON ON THE ULTIMATE TENSILE STRENGTH AND UNIFORM ELONGATION OF VACUUM-MELTED LOW-CARBON PHOSPHORUS STEELS ANNEALED AT 710°C

Figure 4B.

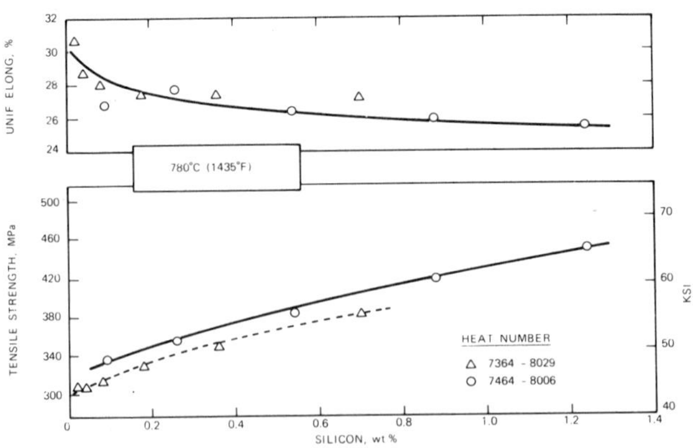

EFFECT OF SILICON ON THE ULTIMATE TENSILE STRENGTH AND UNIFORM ELONGATION OF VACUUM-MELTED LOW-CARBON PHOSPHORUS STEELS ANNEALED AT 780°C

Figure 4C.

Tensile Strength and Ductility of Annealed Strips

The ultimate tensile strength and the ductility (as expressed in percentage of uniform elongation) of the annealed strips are shown in Figures 4A, 4B, and 4C as a function of silicon concentration. As expected, the ductility of the steel decreased with increasing silicon content, whereas the tensile strength increased. However, the decrease in ductility was small, about 4 to 5 percent in uniform elongation when the silicon content was increased substantially from 0.02 to over 1.2 percent. The same amount of silicon addition increased the ultimate tensile strength by about 140 MPa (20 ksi). The effect of annealing temperature on the change in uniform elongation was negligible. For the steel having the maximum silicon content (1.24% Si, Steel E in Table II), and annealed at the lowest temperature (650°C), a tensile strength of about 470 MPa (nearly 70 ksi) was obtained (Figure 4A).

A small but consistent difference in the tensile strengths of the steels made from the two separate heats was shown in Figures 4A, 4B, and 4C. The difference, varying from 10 to 20 MPa (1.5 to 3 ksi), and depending on the annealing temperature, was probably due to a difference in the base composition of the steels, mainly in the carbon and phosphorus contents.

Grain Size of the Annealed Strips

The effects of silicon content and annealing temperature on the grain size of the steels are shown in Figure 5. Although there was some scatter in the data, which became greater with

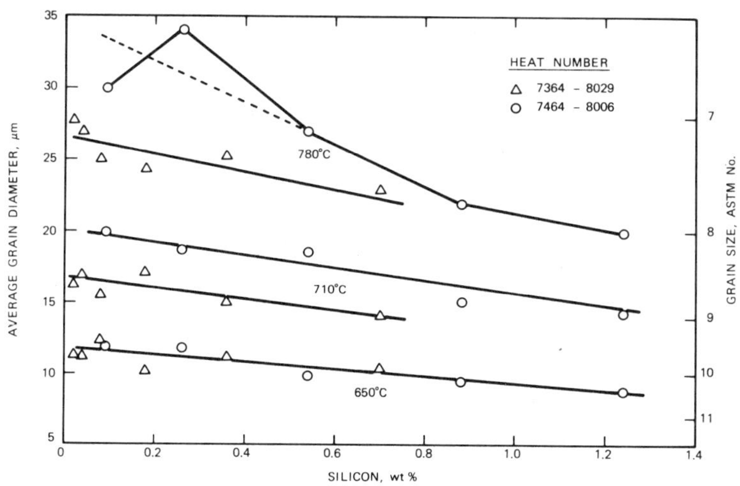

EFFECT OF SILICON CONTENT ON THE GRAIN SIZE OF VACUUM-MELTED LOW-CARBON PHOSPHORUS STEELS ANNEALED AT VARIOUS TEMPERATURES

Figure 5.

increasing temperatures of annealing, the average grain size decreased with increasing silicon content. This can be understood on the basis that, with increasing silicon content, the cold-rolled steel recrystallizes with somewhat greater difficulty. Hence, during the same simulated box-annealing cycle, grain growth occurs to a lesser extent in a steel with high silicon content than in a steel with a low silicon content.

It is interesting to note in Figure 5 that although annealing at 650°C produced no distinction between the grain sizes of two heats of steel (Tables I and II), annealing at 710°C showed clear differences between the grain sizes of the two heats. An even greater separation between the grain-size data of the two heats was indicated after annealing at 780°C, although the scatter of data was also large. After annealing at 780°C, some grains were appreciably larger than the average ferrite grain size, indicating a tendency for coarsening. These observations suggested that the grain-growth characteristics of the steels made from the two heats had some differences, which might have resulted from the base compositional variations. It is known[10] that silicon tends to increase the carbon concentration in austenite by forcing carbon to migrate away from ferrite. Accordingly, the growth of recrystallized ferrite grains during annealing at intercritical temperatures would be reduced by increasing silicon content because of the effect of silicon-carbon interactions on the mobility of the ferrite grain boundaries.

<u>Correlation Between Yield Stress and Grain Size</u>

As shown in previous investigations,[1-3] the yield stress of the annealed phosphorus steels can generally be correlated with grain size according to a Hall-Petch relation.

$$\sigma_y = \sigma_o + k_y d^{-1/2}$$

where σ_y is the yield stress, d is the grain diameter, σ_o is the intercept, and k_y is the slope of the resulting straight lines.

When the yield-strength and grain-size data for the present steels were similarly plotted, the data points fell over such a wide area that a reasonable correlation could not be produced. However, when the data points were distinguished by the silicon content of the steels, simple straightline relationships between σ_y and $d^{-1/2}$ were immediately obvious, as shown in Figure 6. As indicated in Figure 6, the numbers 1 to 11 attached to the data points represent, in the increasing order, the silicon contents of the steels. For steels of low silicon contents, (steels designated 1 to 7) the data points were clustered within a narrow scatter band of about 40 MPa (5.8 ksi) which could be expressed by a straightline relationship for the mean values.

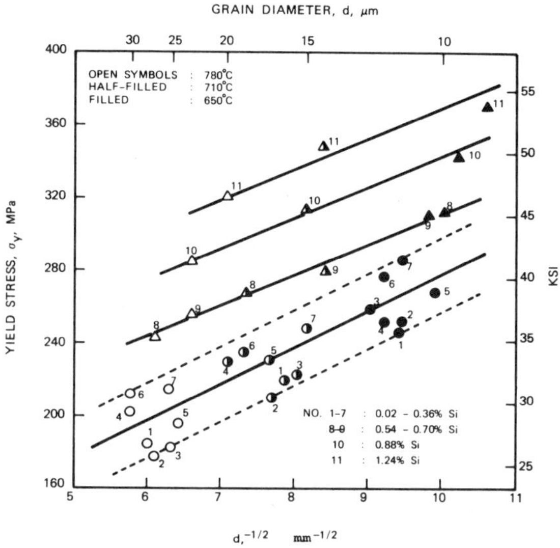

RELATION BETWEEN YIELD STRESS AND GRAIN SIZE OF VACUUM-MELTED LOW-CARBON PHOSPHORUS-SILICON STEELS ANNEALED AT VARIOUS TEMPERATURES

Figure 6.

The steels of higher silicon contents fall into three groups as indicated above and consequently three additional parallel lines can be drawn. The intercepts (σ_o) and the slopes (k_y) of these straightline plots are shown in Table III.

Table III

Hall-Petch Parameters of Plots in Figure 6

Number Designation of Steels	Si, w/o	σ_o, MPa (ksi)	k_y, MPa (ksi)/mm$^{-1/2}$
1 to 7	0.02 to 0.36	76 (11)	20 (2.9)
8 to 9	0.54 to 0.70	144 (21)	17 (2.5)
10	0.88	173 (25)	17 (2.5)
11	1.24	200 (29)	17 (2.5)

These results indicate that σ_o, or the frictional stress, increases with the silicon content of the steel, whereas k_y, the grain-dependence of strengthening rate remains the same or may decrease slightly with increasing silicon content. The values of both these parameters, σ_o and k_y are comparable in magnitude with those reported by Toda et al[11] in phosphorus-aluminum-killed steels, batch- or continuous-annealed, and by Petch[12] for mild steels tested at various temperatures (18, -23, and

-79°C). The physical meaning of these Hall-Petch parameters has been discussed extensively in the literature.[13,14] The observed increase of the σ_o value with increasing silicon can be understood on the basis of lattice friction which increased with increasing solute concentration. The nearly constant, or very slight decrease with increasing silicon, of the k_y value indicates that the strength of the grain boundaries in the present phosphorus-containing steels had not been affected to any appreciable extent by phosphorus-silicon interaction, if any, in the grain boundaries. This observation is consistent with the reported finding that silicon had a weak tendency to combine with phosphorus in iron or steel.[15]

Strain-Hardening Characteristics

The strain-hardening and strain-aging characteristics of the annealed strips were examined by following the same procedures described in earlier investigations.[1-3] The effect of silicon concentration on the strain-hardening exponent, n, of the expression, $\sigma = K\varepsilon^n$, (σ and ε are true stress and true strain, respectively, during homogeneous tensile deformation, and K is a parameter indicating the stress at unit strain) is shown in the upper part of Figure 7. For all annealing temperatures studied

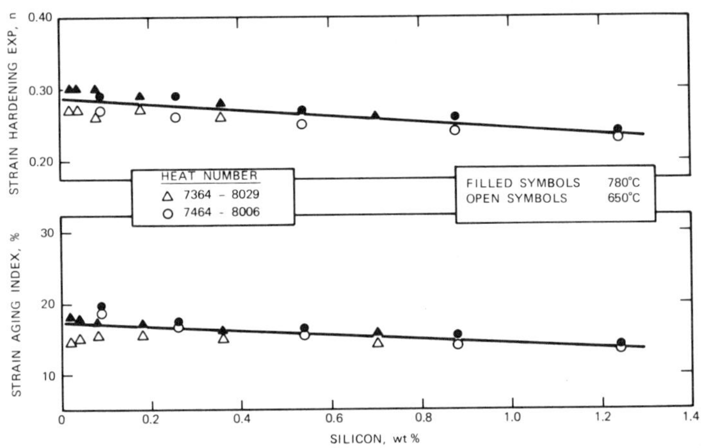

EFFECT OF SILICON ON STRAIN-HARDENING EXPONENT AND STRAIN-AGING INDEX OF VACUUM-MELTED LOW-CARBON PHOSPHORUS STEELS ANNEALED AT VARIOUS TEMPERATURES

Figure 7.

(data for the 710°C anneal were not plotted, because they were intermediate between the data for the 780 and 650°C anneals), the n value decreased with increasing silicon content and with decreasing annealing temperature. However, the variations were all within a narrow range of values, from approximately n = 0.23 to 0.30 for the steel compositions and annealing temperatures studied. These n values were not significantly different from those of as-annealed DQSK aluminum-killed or DQ rimmed steel sheets. The observed dependence of the n value on silicon content and annealing temperature can be explained on the basis of finer grain size or higher strength associated with higher silicon contents, since it is well known that the n value decreases with increasing yield or flow stress.[16)]

Strain-Aging Characteristics

The strain-aging index, normally defined as the percentage increase in the flow stress that results from aging prestrained specimens (strained 8% in tension) for 4 hours at 100°C (212°F) is shown in the lower part of Figure 7. As with the n values, results from strain aging the specimens annealed at 710°C were not presented because they were intermediate between those for the two temperatures shown. In addition to the normal 8 percent prestraining, the data points represented by the open circles were for specimens prestrained only to 5 percent elongation, followed by the same aging treatment and retesting. As can be seen, the difference in the strain-aging index resulting from such a small difference in the amount of prestraining is negligible. The plot indicates that the strain-aging index decreased slightly with increasing silicon concentration, and with decreasing annealing temperature. The reason for these observed effects is at present unclear. The variations in the strain-aging index within the range of silicon additions and annealing temperatures studied were all within a narrow range from about 13 to 19 percent. For rimmed CQ or DQ sheet steels, the strain-aging index is normally around 18 to 23 percent.

Summary and Conclusions

As an extension of a systematic study of the characteristics of phosphorus-containing low-carbon steels, the effects of silicon additions on the annealing texture, plastic anisotropy, and other mechanical properties have been studied. These steels, having a base composition of approximately 0.04C, 0.20Mn, and 0.04 to 0.07P, with various silicon concentrations ranging from 0.02 to 1.24 in weight percent, were vacuum-melted in the laboratory, hot-rolled to 3.8-mm (0.150 in.) thickness, cold-rolled 80 percent, and annealed at 650, 710, and 780°C (1200, 1310, and 1435°F) for 20 hours. The microstructures and crystallographic textures were examined after each processing treatment. The anisotropy parameters and other mechanical properties were determined for the annealed strips by tension tests. The results may be summarized as follows:

1. The additions of silicon increased the yield and tensile strengths of the steel, with only small reductions in ductility. The excellent anisotropy parameters (high r_m and low Δr values) of the phosphorus-containing low-carbon steels were not seriously affected even at relatively high silicon concentrations (up to 0.7%). With further increase in the silicon content (up to over 1.2%), the r_m value decreased only gradually. Thus, by increasing the silicon content of the phosphorus-containing low-carbon

steels, high-strength sheet with deep-drawing capabilities could be produced by appropriate processing treatments.

2. For the range of silicon contents and annealing temperatures studied, the average plastic strain ratio (r_m value) increased, and the yield strength decreased, with increasing annealing temperature. Thus, by proper selection of the steel composition and the annealing temperature, a wide variety of strength and drawability combinations can be obtained.

3. In agreement with recent findings, the r_m values were found to be strain-dependent, and for the strongly (111)-textured sheet, the r_m values decreased with increasing strain.

4. The strain-hardening exponent, or n in $\sigma = K\varepsilon^n$, of the present steels was found to decrease with increasing silicon content and with decreasing annealing temperature. This could be explained on the basis of finer grain size or higher yield strength associated with higher silicon contents.

5. The strain-aging index decreased slightly with increasing silicon content and with decreasing annealing temperature. The reason for these observed effects is at present unclear.

References

1. H. Hu, Texture Cryst. Sol., 1976, Vol. 2, pp. 113-41.

2. H. Hu, Met. Trans. A, 1977, Vol. 8A, pp. 1567-75.

3. H. Hu, "Effect of Carbon on Annealing Texture, Plastic Anisotropy, and Mechanical Properties of Phosphorus Steels," to be published.

4. H. Hu, "Low-Carbon Phosphorus Steels for High-Strength Cold-Rolled Sheet Applications," unpublished work, 1974, U. S. Steel Research Laboratory.

5. H. Hu, "Effect of High Rates of Heating by Electrical Induction Annealing on the Plastic Anisotropy and Strength of Phosphorus Steel Sheets," unpublished work, 1974, U. S. Steel Research Laboratory.

6. W. C. Leslie, Met. Trans., 1972, Vol. 3, pp. 5-26.

7. H. Hu, Met. Trans. A, 1975, Vol. 6A, pp. 945-47.

8. H. Hu, Met. Trans. A, 1975, Vol. 6A, pp. 2307-09.

9. H. Hu, Met. Trans. A, 1976, Vol. 7A, pp. 905-06.

10. P. Abramowitz and R. A. Moll, Met. Trans., 1970, Vol. 1, pp. 1773-75.

11. K. Toda, H. Gondoh, H. Takechi, and H. Masui, Met. Trans. A, 1976, Vol. 7A, pp. 1629-39.

12. N. J. Petch, Phil. Mag., 1958, Vol. 3, pp. 1089-97.

13. J. C. M. Li and Y. T. Chou, Met. Trans., 1970, Vol. 1, pp. 1145-59.

14. R. W. Armstrong, Can. Met. Quar., 1974, Vol. 13, pp. 187-202.

15. H. Kaneko, T. Nishizawa, and K. Tamaki, J. Inst. Met. Japan, 1965, Vol. 29, pp. 159-65.

16. W. B. Morrison and R. L. Miller, in Ultrafine-Grain Metals, p. 183, Syracuse University Press, Syracuse, N. Y., 1970.

STRENGTH, DUCTILITY AND AGING PROPERTIES OF CONTINUOUSLY-ANNEALED DUAL-PHASE HIGH-STRENGTH SHEET STEELS

K. Nakaoka, K. Araki, and K. Kurihara

Nippon Kokan K.K.
Technical Research Center, Steel Product Laboratory
Kawasaki, Japan

Abstract

The use of continuous annealing line equipped with a water quenching device (NKK CAL process) makes possible an easy and economical production of dual-phase high-strength steel. The basic heat cycle is composed of water quenching from intercritical temperatures and subsequent tempering at low temperatures. The water quenching from intercritical temperatures leads to the formation of ferrite-martensite dual-phase structure. The prime objective of the tempering is to minimize the change in the properties of the products due to quench aging. To assure the high paint bake-hardenability that is a distinctive feature of the steel, it is necessary that a reappearance of yield point elongation is suppressed at room temperature. By adjusting the volume fraction of martensite, f_M, in a range from 0.1 to 0.15, the steel showing high bake-hardening and the retarded reappearance of yield point elongation, together with good ductility, is obtained. As interstitial atoms which contribute to bake-hardenability, carbon is more suitable than nitrogen.

Introduction

Dual-phase steels composed of ferrite matrix and martensite have drawn attention as new high-strength materials for automotive applications because of low production cost and superior ductility compared with conventional HSLA steels (1). Moreover the steels have the characteristic that their yield stress increases during the paint-baking treatment at the automotive painting line (bake-hardenability), although they show retarded reappearance of yield point elongation in strain aging at room temperature.

We (2, 3) and D. J. Bailey (4, 5) developed a method to produce dual-phase steels by means of water quenching a low-carbon steel from the intercritical temperature range and subsequently tempering at low temperatures (200-300°C) for a short time. This method has been industrialized using a continuous annealing line equipped with water quenching apparatus (NKK CAL process) (6). This method involves extremely low production costs and facilitates the production of steels with superior ductility and high bake-hardenability through control of chemistry and heat cycle.

In general, the properties of dual-phase metals are strongly affected by the properties, volume fraction, distribution and morphology of their constituent phases (7). Therefore, the stress-strain relationship of dual-phase steels should be influenced by the work hardening behaviors of the ferrite matrix and the martensite, and the volume fraction and morphology of martensite. One of the purposes of this paper is to describe our recent point of view on the strength and ductility of the dual-phase steels. Another purpose is to make clear the role of interstitial atoms and the harder phase on the characteristic aging behaviors of the dual-phase steel.

Strength and Ductility

In a recent paper (8), we derived a theoretical equation, on the basis of continuum mechanics, which describes the overall stress-strain curve of dual-phase steel by using those of its constituent phases. The appropriateness of the theoretical equation was confirmed by comparing the equation with experimental results, and the following conclusions were drawn: 1) The work hardening behavior of dual-phase steel is most strongly affected by that of ferrite matrix. Therefore, the prime requirement for improving the ductility of dual-phase steel is to improve the ductility of ferrite matrix. 2) The existence of martensite contributes to the strengthening of the steels without reducing the product of the tensile strength and the elongation (TS. El-value) providing that the martensite volume fraction (f_M) is below 0.2. 3) Although any improvement in the ductility of martensite contributes to improving that of dual-phase steel, this contribution is less than that provided by improving the ductility of the ferrite matrix (in the case of f_M below 0.2).

The above conclusions permit us to explain the ductility of dual-phase steels as follows: The excellent ductility reported for most of the dual-phase steels may be attributed to the fact that all these steels contain considerable amount of solid solution strengthening elements such as Si, P and Mn and/or martensite formation elements. In addition to martensite strengthening (which does not lead to a deterioration in the TS.El-value), solid solution strengthening is preferred as it is known to be a favorable strengthening mechanism in ferritic steels from the viewpoint of ductility. Precipitation strengthening, on the other hand, is not favored because of its detrimental effect on ductility (9).

Experimental Details

In order to confirm the above considerations, we conducted an experiment to investigate the effect of solid solution strengthening elements and f_M on the TS.El-value. Silicon was chosen as the solid solution strengthening element and Si-bearing and Si-free steels were compared. The materials and heat treatment cycles are shown in Figure 1. The steels were cold rolled to 0.8 mm thickness prior to heat treatment. The heat treatment cycles consisted of water quenching from the α-γ temperature range and subsequent low-temperature tempering, and closely approximated the heat treatment cycles for dual-phase steels in the NKK CAL process. The volume fraction of martensite was varied by carbon content and quenching temperature. A heat treatment cycle employing furnace cooling was also used to prepare materials with an f_M of nearly zero (Figure 1). After heat treatment, JIS No. 5 tensile specimens were prepared with their tensile directions perpendicular to the rolling direction and temper rolled by 1 pct. Tensile tests were conducted on an Instron-type tensile machine.

STEELS	CHEMICAL COMPOSITIONS(wt%)			
	C	Si	Mn	P
1 % Si	0.02 ~0.31	0.95 ~1.05	0.26 ~0.39	< 0.012
NO ADDITIVES	0.02 ~0.30	< 0.01	0.25 ~0.38	< 0.010

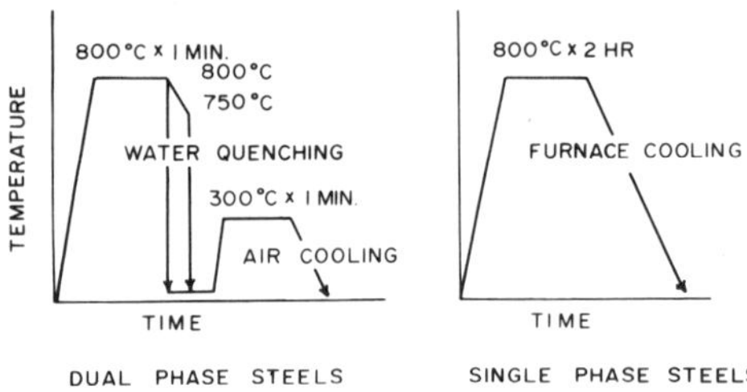

Fig. 1 - Materials and heat treatment cycles employed for investigating ductility.

Results and Discussion

Results are shown in Figure 2. We can recognize the following four facts which confirm the preceding considerations on the ductility of dual-phase steels: 1) The TS.El-values of the Si-bearing steels are superior to those of the Si-free steels for the whole f_M range. 2) The degree of superiority can be accounted for by the difference in the TS.El-values of the two steels which do not contain martensite. 3) In the f_M range below 0.2, the TS.El-values for both types of steel remain constant and independent of f_M. 4) The existence of martensite greatly contributes to the strength of the steels.

The first two facts prove that the superior ductility of steels containing the solid solution strengthening element, Si, results from an improvement in the TS.El-value of the ferrite matrix, the effect of other metallurgical factors being small. Recently, the existance of retained austenite in the martensite phase of a Si-bearing dual-phase steel has been recognized (10). In our theoretical treatment (8), the effect of the retained austenite can be discussed in terms of the change in the properties of the second phase. The theoretical conclusion was that if the work hardening coefficient of the second phase increases to a certain extent the TS.El-value is improved considerably. So, if retained austenite is present and is improving the work hardening coefficient, the increase in the TS.El-value can be attributed to the existence of the retained austenite. From our knowledge of the TRIP phenomenon, an improvement in the work hardening coefficient is quite probable. In the present experiment, however, no evidence of the presence of retained austenite was recognized. And it should be pointed out again that the favorable influence of Si addition was demonstrated in the single phase samples. Our theoretical treatment shows that this ductility improvement of the ferrite matrix plays the most important role in the improvement of the TS.El-value of dual-phase material. Therefore, at least in our case, the beneficial effect of the Si addition is attributed to the ductility improvement of the ferrite matrix.

Our theoretical treatment (8) did not take the effect of the morphology of martensite into account. From the considerations on the basis of a continuum mechanics model, the strengthening ability of plate-like or needle-like martensite phase will be higher than that of a spherical martensite phase. However, these changes in morphology should not have an influence on the TS.El-value.

The deterioration in TS.El-value for f_M over 0.2 in both types of steels is considered to be caused by the rapid onset of void formation in the close vicinity of the martensite particles prior to the occurrence of plastic instability (8).

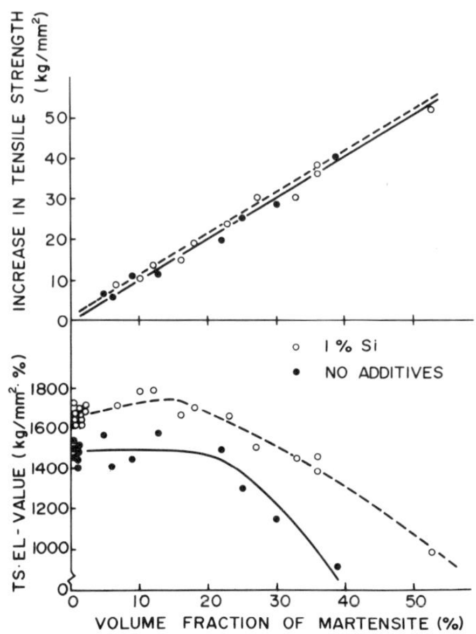

Fig. 2 - Influence of volume fraction of martensite on the change in tensile strength and tensile strength.elongation value in Si-bearing and Si-free steels.

Aging Properties

The significant feature of the water quenching and subsequent tempering is the capability of assuring bake-hardenability by suppressing the room temperature recovery of yield point elongation (11, 12). In this paper, the paint baking treatment (BH treatment) is simulated by an aging treatment at 170°C for 20 minutes, and the resultant increase in yield stress is termed "bake-hardenability". Room temperature aging is assessed by the reappearance of yield point elongation after aging at 38°C for 30 days.

Effect of Heat Cycle on Bake-Hardenability

Figure 3 shows the effect of quenching and tempering temperatures (tempering time 1 minute) on bake-hardenability of a 0.07% C aluminum-killed steel for commercial sheet steel. Figures attached to several data points indicate solute carbon contents (ppm) determined by internal friction.

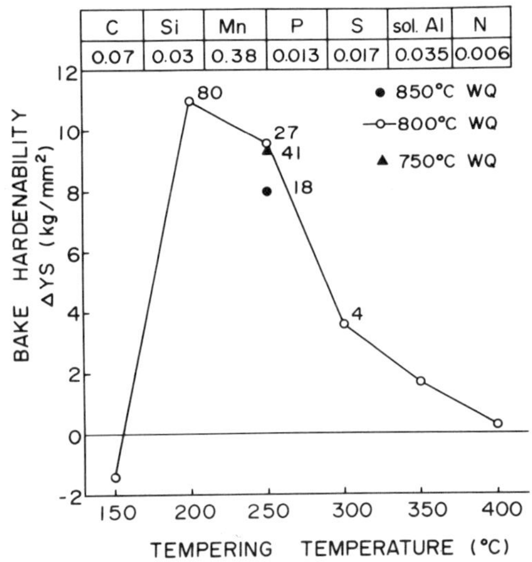

Fig. 3 - Influence of heat cycle on the bake-hardenability of dual-phase steels produced from a low-carbon Al-killed steel.

The bake-hardenability is strongly affected by tempering temperature. When the temperature is over 200°C, the bake-hardenability is high and decreases with increasing temperature and reduces to zero at 400°C. This decrease of solute carbon content and the corresponding drop in strength indicates that the bake-hardening is a type of strain aging. Below 200°C, however, the bake-hardenability decreases in spite of the increase in solute carbon content. The interesting fact is that in materials that are either as-quenched (3) or insufficiently tempered (e.g., 150°C for 1 minute) prior to bake-hardening, further softening takes place during the BH treatment (13). The effect of quenching temperature (Figure 3) in the range 750-850°C on bake-hardenability is rather small and provides for an adequate level of solute carbon content (20-40 ppm at 250°C tempering). In industrial practice, the quenching temperature is determined from the viewpoint of strength and ductility without regard to the bake-hardenability; on the other hand, the tempering temperature is determined with regard to the bake-hardenability.

Figure 4 shows the relationship between bake-hardenability and the amount of interstitial solute atoms for the water quenched and tempered dual-phase steels. It can be seen that at least 20 ppm of solute carbon is necessary for a bake-hardenability of around 10 kg/mm^2.

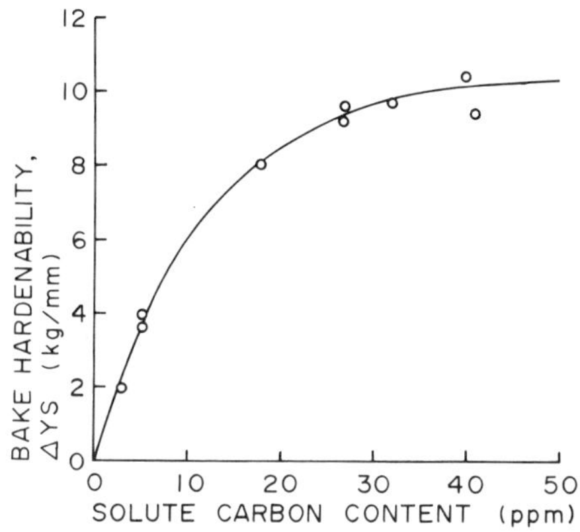

Fig. 4 - Relation between the amount of solute carbon and bake-hardenability.

Effect of Martensite and Interstitials on Aging Behavior

The aging behavior of dual-phase steel composed of ferrite and martensite is supposed to be influenced by the volume fraction of second phase (f_M) as well as by the amount of interstitials (14). In this research, the volume fraction of martensite was varied by changing the carbon content and quenching temperature (750°C, 800°C and 850°C). Experiments were undertaken for two cases; one where the interstitial is mainly carbon, and the other where the interstitial is mainly nitrogen. For the solute carbon case, aluminum-killed steels were utilized; and for the solute nitrogen case, capped steels were utilized with an overaging treatment at 300°C to allow full carbon precipitation.

The aging behaviors studied were the bake-hardenability and also the reappearance of yield point elongation after storage at 38°C for 30 days.

Solute Carbon Case: Figure 5 shows the experimental results for aluminum-killed steels with carbon contents of 0.03 to 0.16%. The composition range of these steels is shown in Table I. In this case, the interstitial atoms are largely carbon (because of the AlN precipitation during the heating prior to quenching) and their amount is estimated to be in the range from 20 to 40 ppm according to Figure 3.

As shown in Figure 5, the bake-hardenability appears to be suppressed in the range where f_M exceeds 30 to 40%. This suppression is not a serious problem because in industrial practice, the f_M is controlled to under about

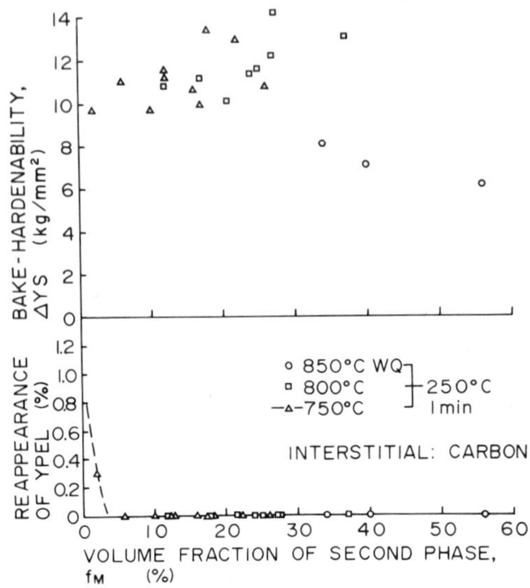

Fig. 5 - Influence of the volume fraction of second phase on the aging behavior of a low-carbon Al-killed steel (solute-carbon case).

Table I. Chemical Composition of Aluminum-Killed Steels

C	Si	Mn	P	S	sol.Al	N
0.03 / 0.16	0.02 / 0.07	0.32 / 0.73	0.010 / 0.014	0.007 / 0.024	0.040 / 0.055	0.0039 / 0.0070

0.2 (20%) to avoid impairing the TS.El-value as described in the preceding section. In this f_M range (under about 20%), bake-hardenability is not influenced by f_M, and at the same time the influence of quenching temperature is not observed.

The reappearance of the yield point elongation (Figure 5), which is a measure of room temperature aging behavior, is not observed, the exception being the specimen whose f_M is too small to be included in practical dual-phase steels.

It can be deduced from these relationships that the f_M which allows the coexistence of bake-hardenability of over 10 kg/mm^2 and complete suppression of yield point elongation recovery at room temperature, lies in the range from 4% to 30%.

Solute Nitrogen Case: Figure 6 shows the results for capped steels with carbon contents of 0.03% to 0.19%. The chemical composition range of these steels is listed in Table II. For comparison, some of the specimens were not water-quenched but air-cooled (AC) to room temperature to obtain microstructures with pearlite as the second phase (solid circle in Figure 6). All the specimens were subsequently tempered under the same conditions.

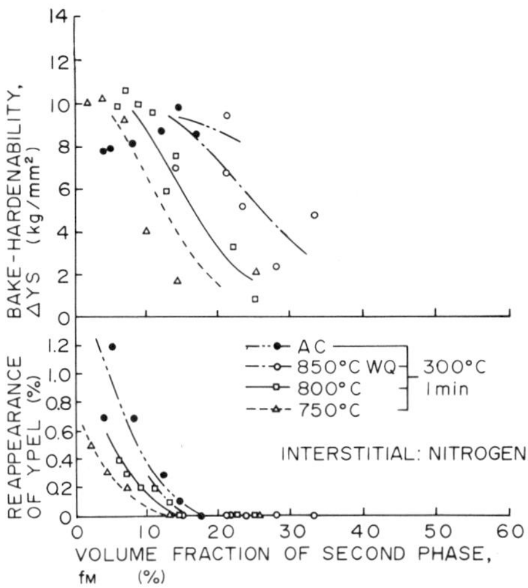

Fig. 6 - Influence of the volume fraction of second phase on the aging behavior of a capped steel (solute-nitrogen case).

Table II. Chemical Composition of Capped Steels

C	Si	Mn	P	S	sol.Al	N
0.03	trace	0.27	0.007	0.005	trace	0.0012
0.19	0.01	0.47	0.029	0.025	0.001	0.0022

The amount of interstitial atoms was measured to be about 15 ppm and was independent of heat cycle. Because the solute carbon was decreased to a low level (around 4 ppm) by tempering at 300°C (Figure 3), the interstitial atom in this case is largely nitrogen.

As Figure 6 shows, the bake-hardenability is suppressed when f_M exceeds 10 or 20%, the suppression being greater with increasing f_M. The reappearance of yield point elongation behavior, which is a measure of room temperature aging, does not disappear until f_M reaches a level of about 15%.

In contrast to the solute carbon case, the steels with largely solute nitrogen instead of solute carbon give a narrow range of f_M where the combination of sufficient bake-hardenability and no reappearance of yield point elongation is assured. Another distinct feature of the nitrogen case is that there is an influence of quenching temperature. As can be seen in Figure 6, an increase in the quenching temperature leads to a weaker suppression of room temperature aging, i.e., the reappearance of the yield point elongation. This influence is discussed in a following section (Difference in Aging Behavior Between Solute-Carbon Material and Solute-Nitrogen Material).

Discussion

The Effect of the Second Phase in Suppressing Aging Behavior: In the experimental results as shown in Figure 5 and Figure 6, it was recognized that both the bake-hardenability and the room temperature aging behavior are suppressed by increasing the volume fraction of second phase, f_M. These phenomena are considered to be caused by an effect in which each neighboring region of second phase becomes the origin of a Lüders front which removes the yield point elongation from the stress-strain curve. This effect is considered to be the result of a disturbance (deformation) of the ferrite matrix by the martensitic transformation of the second phase (15). Even if the ferrite matrix is aged enough to cause a reappearance of the yield point elongation, it is believed that with a sufficient level of f_M, the stress-strain curve of the dual-phase materials will show no yield point elongation.

Based on the law of mixtures, the stress-strain curve of a dual-phase steel can be expressed by

$$\sigma = (1 - f_M)\ \sigma_F + f_M\ \sigma_M; \qquad [1]$$

where, σ, σ_F and σ_M are the stress-strain curves (i.e., the stresses at a given unit strain) of dual-phase steel, ferrite and martensite, respectively, and f_M is volume fraction of martensite. In this formula, the degree of disturbance as mentioned above is expressed through the second term, $f_M\ \sigma_M$, and increases with increasing $f_M\ \sigma_M$. The effect of quenching temperature recognized in Figure 6 can be associated with this effect, since a higher quenching temperature decreases σ_M.

As a measure of the degree of disturbance, taking σ_M into account, let us use the following index:

$$f_M\ \sigma_M/[(1 - f_M)\ \sigma_F + f_M\ \sigma_M]. \qquad [2]$$

This index expresses the ratio of the stress caused by the martensite to the whole stress. The data of Figure 6 are replotted as a function of this index in Figure 7. In the index, σ_M/σ_F was assumed to be 6, 4.5 and 3 for the quenching temperatures of 750°C, 800°C and 850°C, respectively, as calculated on the basis of the martensite hardness corresponding to a carbon content (16) estimated from the equilibrium phase diagram (17). For the specimen with pearlite as the second phase, σ_M/σ_F was assumed to be 2.5.

As shown in Figure 7, a better correlation is obtained between the experimental results and the above-mentioned index than with f_M (Figure 6). The above consideration seems to provide a reasonable indication that in addition to the volume fraction of martensite, the properties of martensite (σ_M) also influence the aging behavior of dual-phase steels.

Unlike Figure 6, however, Figure 5 which was for the solute carbon case shows no influence of quenching temperature, that means, no influence of σ_M. This implies that the mechanism of bake-hardenability in the solute carbon case differs somewhat from that in the solute nitrogen case.

Mechanism of Bake-Hardening: To clarify the aging mechanism during bake-hardening, aging behaviors were investigated at 65°C, 80°C, 100°C,

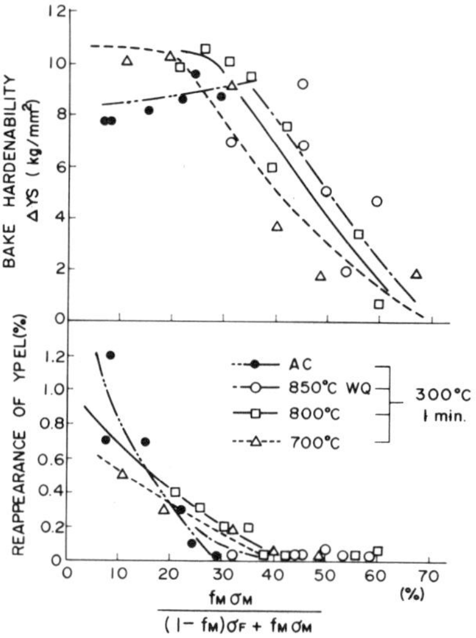

Fig. 7 - Rearrangement of Figure 6 by using the index $f_M \sigma_M / [(1 - f_M) \sigma_F + f_M \sigma_M]$ instead of f_M.

120°C, 140°C and 170°C, and the activation energy was estimated by means of the Arrhenius plot (18). Two kinds of samples, the compositions of which are listed in Table III, were subjected to the experiment. Specimen identification, heat cycle, type and amount of interstitials, and volume fraction of martensite are listed in Table IV. Samples C and N are solute-carbon and solute-nitrogen materials, respectively, and are further identified by a number which refers to the percent volume fraction of martensite in the material.

Table III. Chemical Composition of Aluminum-Killed Steels (C0, C11) and Renitrogenized Steels (N0, N9)

Specimen	C	Si	Mn	P	S	sol.Al	N
C0, C11	0.06	0.02	0.32	0.012	0.017	0.054	0.0044
N0, N9	0.06	trace	0.33	0.008	0.015	0.001	0.0072

Figure 8 is an Arrhenius plot of the temperature and times required to produce an increase in yield stress of 5 kg/mm^2. It can be seen that aging proceeds more slowly in the solute-carbon samples than in the solute-nitrogen samples. However, the activation energies of the two samples show no appreciable difference and lie in the range between 25 and 30 kcal/mole. This value is different from that associated with strain aging which is reported to be from 18 to 20 kcal/mole and equal to that of diffusion of carbon or nitrogen atoms in ferrite (19). Therefore the mechanism of the present bake-hardening phenomenon does not seem to be merely the diffusion of carbon or nitrogen in ferrite.

Table IV. Heat Cycle and Volume Fraction of Martensite and Amount of Interstitials

Specimen	Heat Cycle		Volume Fraction of Martensite (%)	Amount of Solute Atom (ppm)
CO	700°C WQ	250°C x 1 min AC	0	Carbon 30
C11	800°C WQ		11	
NO	700°C WQ	300°C x 10 min FC	0	Nitrogen 30
N9	800°C WQ		9	

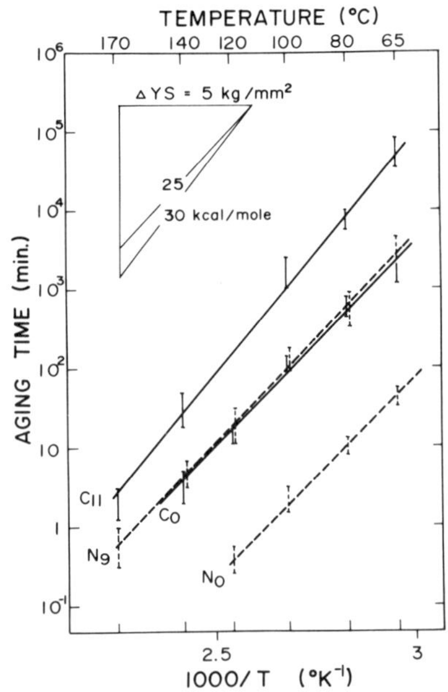

Fig. 8 - Arrhenius plots for the aging behavior of solute-carbon type and solute-nitrogen type dual-phase steels (C11, N9), and corresponding single phase steels (C0, N0).

In order to provide a more fundamental understanding of the behavior of interstitials during bake-hardening, a test was planned in which samples were subjected to three similar consecutive cycles of 1% temper-rolling (TR) and subsequent BH treatment (170°C - 20 minutes). Sample C11 (solute-carbon type) and sample N9 (solute-nitrogen type) in Table IV were subjected to the experiment. Yield stress and Snoek peak were measured after every temper-rolling and BH treatment cycle.

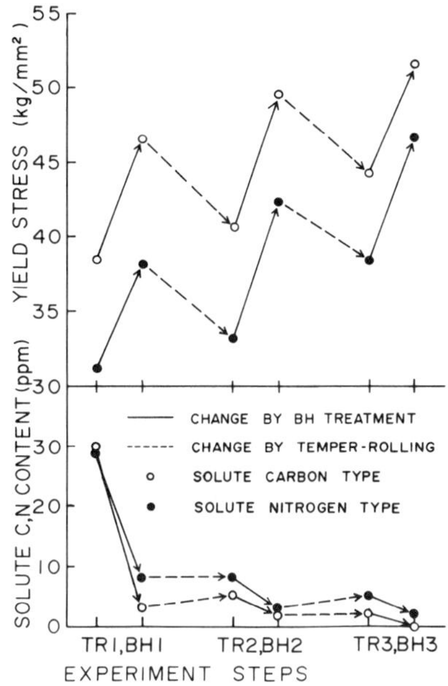

Fig. 9 - Variations of the yield stress and amount of interstitials through repeated cycles of temper-rolling (TR) and bake-hardening (BH).

The results are shown in Figure 9. The condition TR1 corresponds to the ordinary as-shipped product. In this stage, both of the samples had about 30 ppm of interstitial atoms. After the first BH treatment, BH1, the yield stress increased as expected and the interstitial solute decreased to below 10 ppm. After subsequent temper-rolling (TR2), the yield stress decreased by several kg/mm^2 and the interstitial solute remained at the lower levels.

It is very interesting to note that even at this stage, TR2, the steels had a bake-hardenability equivalent to the as-shipped product, TR1, as observed in the difference of yield stress between conditions TR2 and BH2. The corresponding data on the interstitial content show that only 2 or 3 ppm of interstitial atoms were responsible for the second bake-hardenability. Almost the same tendencies were observed for the TR3 and BH3 conditions.

These curious phenomena that an interstitial content of over 20 ppm is necessary at the first BH treatment but only a small amount plays a role in the subsequent temper-rolling and BH treatments, imply that the bake-hardenability is also invoked by something other than mere migration of solute carbon or nitrogen to dislocations. The bake-hardenability of samples after the second or third temper-rolling is considered to be associated with interstitial atoms trapped near the dislocations introduced during temper-rolling. There is a possibility that the rearrangement of such trapped atoms might occur near the newly introduced dislocations. Thus, the mobile dislocations are locked by these atoms and produce the yield stress increase.

If this rearrangement mechanism is responsible for the bake-hardenability of samples after the TR2 and TR3 treatments, there is no reason to deny the possibility of this mechanism in the case of the TR1 samples. In the first BH treatment, i.e., the usual one, the rearrangement mechanism of carbon or nitrogen atoms near the dislocations might well take place after the Cottrel atmosphere is formed by the usual diffusion process. It seems quite probable that the controlling process of the BH phenomenon is not solely the diffusion of interstitials to dislocations but the rearrangement of carbon or nitrogen atoms once trapped by the dislocations.

According to this assumption, the activation energy of the BH phenomenon can be accounted for by the sum of the interaction energy of dislocations and interstitial atoms, and the activation energy of lattice diffusion of interstitial atoms. This agrees with the result from Figure 8, that the activation energy was greater than that for carbon or nitrogen diffusion.

As to the augmentation of activation energy by the interaction with dislocations, there have been papers dealing with the Köster peak in internal frication (20, 21) or serrated flow stress in the tensile test (22) which give activation energies of 30-35 kcal/mole. Although these values are somewhat higher than observed here, they do support the plausibility of the above assumption.

Difference in Aging Behavior Between Solute-Carbon Material and Solute-Nitrogen Material: It is clear from Figures 5 and 6 that the suppression of bake-hardenability as the volume fraction of martensite increases is weaker in the solute-carbon type dual-phase steel than in the solute-nitrogen type. This seems to imply that the anchoring of dislocations after the BH treatment is stronger in the solute-nitrogen type material.

In order to examine the degress of the anchoring effect, experiments similar to that shown in Figure 9 were carried out involving both solute-carbon type and solute-nitrogen type materials. Two similar consecutive temper-rolling and subsequent aging cycles were applied to the specimens and the yield stress was measured after every temper-rolling and aging treatment. In the first cycle, the initial aging was carried out at 200°C for 1 hour with the second aging being the BH treatment. In the second cycle, the temperature of the initial aging was raised to 250°C, the second aging being also BH treatment.

The results of these experiments are summarized in Figure 10. The yield stress increase produced by the initial aging is similar to that observed after the first BH treatment in Figure 9, being a little larger in the first cycle because of the promoted aging condition and slightly smaller in the second cycle because of over-aging.

In contrast to the results in Figure 9, the solute-carbon type sample in Figure 10 shows very little bake-hardenability after the second temper-rolling. Furthermore, the bake-hardenability appears to be reduced as the initial aging temperature is raised from 200°C to 250°C. However, the solute-nitrogen type sample shows a considerable degree of bake-hardenability after the second temper-rolling even though the first aging temperature was raised to 200°C or 250°C.

The above results indicate that when the initial aging teatment is enhanced by adopting higher temperatures and longer aging times, the potential carbon for subsequent bake-hardenability is appreciably reduced, while the potential nitrogen can survive the treatment. Qualitatively speaking, the above results show that aging proceeds to a further extent in the solute-

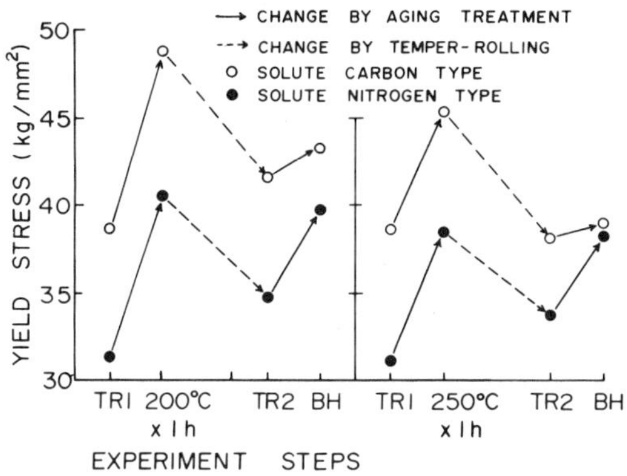

Fig. 10 - Effect of prolonged initial aging treatment on a subsequent bake-hardenability treatment.

carbon sample than in the solute-nitrogen sample. Hence a stronger resistance to the bake-hardenability suppression by the martensite can be expected in the solute-carbon sample.

The difference in the aging enhancement of the solute-carbon and solute-nitrogen samples can be attributed to the difference in the ratio of solute amount to solubility: the solute-carbon sample con tained more solute carbon than solubility, while in the solute-nitrogen sample amount of solute nitrogen was within the solubility limit.

Contrary to the case of bake-hardenability, it can be seen in Figures 5 and 6 that the solute-carbon sample exhibits a stronger suppression of the recovery of the yield point elongation at 38°C than the solute-nitrogen sample does. This seems to be associated with the tendency for aging to proceed slower in the solute-carbon sample than in the solute-nitrogen sample as shown in Figure 8. When the aging temperature is as low as 38°C, the aging of the solute-carbon sample is considered to be in an earlier stage, while that of the solute-nitrogen sample is assumed to be in a stage close to the bake-hardened state. Hence the suppression of the reappearance of the yield point elongation at near room temperature is stronger in the solute-carbon material than in the solute-nitrogen material.

Summary and Conclusions

Studies were made on strength, ductility and aging properties of water-quenched and subsequently tempered dual-phase steels.

In Si-bearing and Si-free steels, effect of the volume fraction of martensite on the strength and ductility relationship was investigated and the following conclusions drawn:

1. Tensile strength increased linearly with the volume fraction increase of martensite.

2. The product of tensile strength and elongation remains unchanged up to 15-20% volume fraction, but decreases monotonically with further increases in the volume fraction of martensite.

3. The Si-addition improves the product of tensile strength and elongation over the whole range of the martensite volume fraction investigated; the favorable influence of Si was also observed in single phase samples.

4. The above results can be accounted for by a theoretical treatment in the authors' recent paper (8).

The effect of interstitial content and martensite volume fraction on the aging behavior of dual-phase steels was investigated with the following conclusions:

5. About 20 ppm of interstitial solute atoms is necessary to provide a bake-hardenability of around 10 kg/mm^2.

6. The presence of a harder second phase (martensite) suppresses the yield stress increase and reappearance of yield point elongation.

7. As an interstitial atom, carbon is more beneficial than nitrogen because it gives wider range of martensite volume fraction where a combination of high bake-hardenability and the complete suppression of the recovery of yield point elongation is assured.

8. The activation energy for bake-hardening determined to be around 30 kcal/mole. This suggests that the bake-hardening mechanism involves the rearrangement of interstitial atoms trapped near the dislocations.

9. Aging proceeds faster in the solute-nitrogen material than in the solute-carbon material. However the locking of the dislocations after the enhancement of the aging is weaker in the solute-nitrogen material than in the solute-carbon material.

References

1. K. Araki, K. Nakaoka, M. Abe and N. Ohashi; Proceedings of 9th Biennial Conference of IDDRG, (1976), 39.

2. K. Uchida and K. Araki; U.S. Patent 3904446, (1975).

3. K. Araki, S. Fukunaka and K. Uchida; Trans. ISIJ, 17, (1977), 701.

4. D. J. Bailey; U.S. Patent 3930907, (1976).

5. D. J. Bailey; SAE Paper 760715, (1976).

6. H. Naemura, Y. Fukuoka, S. Osaka and H. Ishioka; Nippon Kokan Technical Report No. 24.

7. For example, H. Fischmeister and B. Karlsson; Z. Metallkde, 68, (1977), 311.

8. K. Araki, Y. Takada and K. Nakaoka; Trans. ISIJ, 17, (1977), 710.

9. H. Masui and H. Takechi; Trans. ISIJ, 16, (1976), 69.

10. J. Y. Koo and G. Thomas; Mater. Sci. Eng., 24, (1976), 187.

11. K. Uchida, K. Araki, S. Fukunaka, Y. Koike and T. Kurihara; Tetsu-to-Hagane, 61, (1975), S149.

12. K. Nakaoka, K. Araki, K. Iwase, K. Uchida and S. Fukunaka; ibid., 61, (1975), S150.

13. K. Nakaoka, K. Araki, K. Uchida, Y. Takada and S. Fukunaka; ibid., 61, (1975), S570.

14. K. Nakaoka, K. Araki, K. Kurihara and S. Fukunaka; ibid., 61, (1975), S571.

15. K. Nakaoka, K. Araki and O. Kimpara; Sosei-to-Kako, 17, (1976), 875.

16. Japan Institute of Metals; Kinzoku Binran, Maruzen Co., Tokyo, 1961.

17. Metals Progress; Data Book, (1973) mid-June, 143.

18. K. Kurihara, K. Iwase, K. Araki and K. Nakaoka; Tetsu-to-Hagane, 62, (1976), S169.

19. S. Harper; Phys. Rev. 83, (1951), 709.

20. W. Köster and G. Kampschulte; Arch. Eisenhuttenwes., 32, (1961), 809.

21. H. Ino and T. Sugeno; Acta Met., 15, (1967), 1197.

22. P. G. Forrest; Proc. Roy. Soc., A-242, (1957), 223.

STRUCTURE-PROPERTY RELATIONSHIPS FOR VAN-QN DUAL-PHASE STEELS

J. H. Bucher and E. G. Hamburg

Jones and Laughlin Steel Corporation
Graham Laboratory
Pittsburgh, Pennsylvania

VAN-QN is a dual-phase steel containing a controlled amount of martensite and/or lower bainite in a ferritic matrix to develop the desired combination of strength and formability. The high work hardening rate of this steel permits the development of high strengths in formed parts in combination with excellent formability. The microstructural features controlling the strength-formability balance are described, and compared with similar relationships existing for conventional HSLA steels. VAN-QN with the desired combination of strength and formability has been produced by heat treatment of cold rolled and hot rolled steels. Moreover, a galvanized product has been made. The potential usefulness of higher and lower strength versions of the current commercial product is discussed.

Introduction

Within the context of this paper dual-phase steels are defined as those steels whose microstructure primarily consists of a mixture of ferrite and martensite or lower bainite. There is considerable literature describing the metallurgy of these steels, for example references 1-5. Moreover, the many and varied papers included in this symposium will be a significant addition to this literature. The possible role of retained austenite in the development of the desired properties is highlighted in several contributions. Earlier this year a paper was presented which described the characteristics of Jones & Laughlin's proprietary product in this category, VAN-QN™ (6). Since that time, thinner gauge cold rolled material and galvanized steel have been produced. This paper will put the strength and formability characteristics of these materials in perspective with earlier data. The utilization of dual-phase steels in cold rolled gauges may be a significant part in the necessary weight reduction of vehicles. The information to be presented is directed toward answering the question of the value of such materials.

The data describing production materials will be analyzed from the point of view of existing theories of dual-phase materials. Also, data obtained from laboratory experiments will be used to extend the information obtained from these production materials. To date, commercial utilization of these materials has largely centered within a relatively limited range of gauges and strengths (7). It will be shown that a very significant potential exists for steels with lower strength and better formability as well as higher strength and somewhat restricted formability in comparison with materials in use today.

Thin Gauge Cold Rolled and Galvanized Products

The steel composition of VAN-QN falls within the basic VAN-80* composition given in Table I. To achieve the desired combination of strength and formability, the composition is controlled and restricted consistent with the properties desired and the characteristics of the production equipment utilized. In all cases, the hot rolled or cold rolled steels were given carefully controlled heat treatments using existing continuous annealing facilities. For commercial reasons, specific descriptions of alloy composition and the heat treatment cycles will not be made. Rather the properties obtained and obtainable will be described in detail. Essentially, the data presented will characterize these materials in terms of strength-formability combinations.

TABLE I

CHEMICAL SPECIFICATION OF VAN-80

	% C	% Mn	% Si	% S	% P	% Al	% V	% N
VAN-80	.18 max.	1.60 max.	.60 max.	.015 max.	.04 max.	.02 min.	.05 min.	.005 min.

* VAN-80 is a registered trademark of the Jones & Laughlin Steel Corporation.

Figure 1 is a plot of the uniform elongation as a function of tensile strength for various materials. These data are descriptive of cold rolled steels as thin as .022" and hot rolled steels as thick as .165". Also included are data describing galvanized steels in both cold rolled and hot rolled gauges. Uniform elongation was selected rather than total elongation as a measure of ductility because the former property is essentially insensitive to the effect of gauge. Moreover, the uniform elongation is a well known indicator of the stretchability of steel. Figure 1 shows a significant effect of strength on ductility. Moreover, at a given strength level there is a significant variation in ductility dependent upon chemistry and processing combinations. However, the strength-ductility combination is essentially independent of gauge and steel condition, that is, cold rolled, hot rolled or galvanized. Therefore, previous descriptions of the characteristics of dual-phase materials in general or VAN-QN in particular which were based largely upon the properties of hot rolled steels can be generalized to include steels in the cold rolled and galvanized conditions.

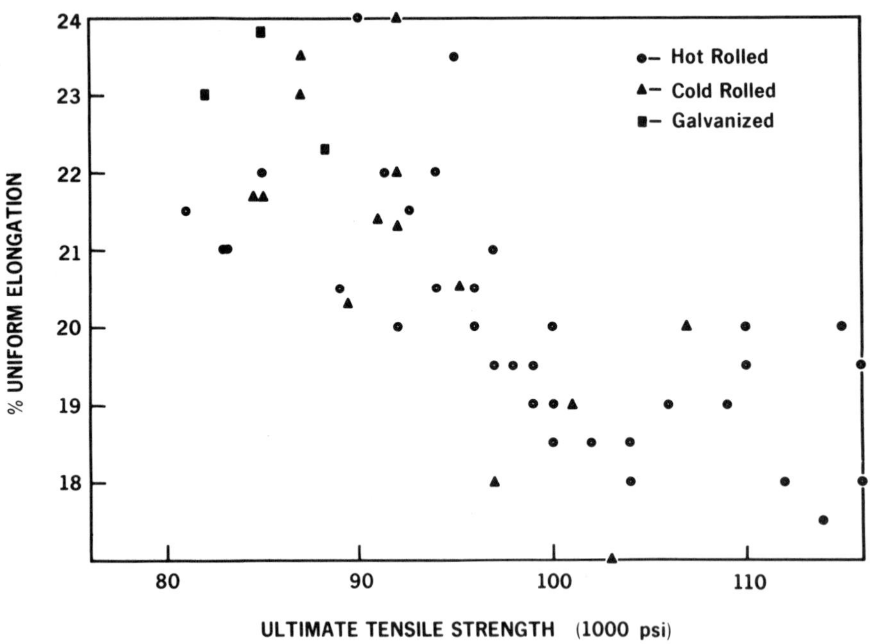

Figure 1. Uniform elongation-strength combinations are seen to be essentially the same for cold rolled, galvanized and hot rolled VAN-QN.

Table II displays the directional tensile properties of a representative hot rolled and cold rolled VAN-QN material. For all intents and purposes the properties are seen to be the same. The lower values of total elongation for the cold rolled steel are simply the effect of the thinner gauge of this material versus the hot rolled steel. Figure 2 is a plot of the relationship of dome depth in a standard stretchability test as a function of strength level. Once again, the relationship is seen to be the same for VAN-QN

TABLE II

TYPICAL TENSILE PROPERTIES OF VAN-QN PRODUCED FROM HOT ROLLED (~.1" Thick) AND COLD ROLLED (~.03" Thick) STEELS ARE COMPARED

Specimen	Orientation	Yield Strength (1000 psi)	Tensile Strength (1000 psi)	% Uniform Elongation	% Total Elongation	n-value	r-value
Hot Rolled	Longitudinal	54.2	96.1	22.0	32.5	.198	.81
	Transverse	51.1	95.6	23.5	32.0	.212	.93
	Diagonal	53.9	96.9	22.0	31.0	.198	.86
	Average	53.3	96.4	22.4	31.6	.202	.87
Cold Rolled	Longitudinal	45.8	92.3	22.8	25.6	.205	.87
	Transverse	47.4	95.3	21.4	24.0	.194	.96
	Diagonal	46.9	94.4	22.5	25.6	.203	.92
	Average	46.8	94.1	22.3	25.2	.201	.92

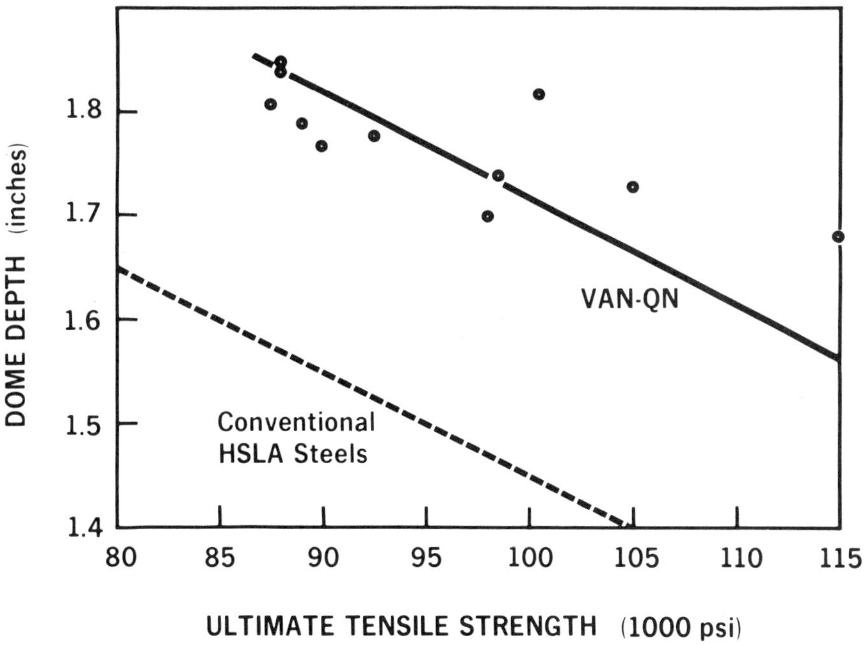

Figure 2. Superior stretchability is indicated for VAN-QN independent of gauge and whether produced from hot rolled or cold rolled steels.

independent of the specific material type. The formability is seen to be significantly better than that of conventional high strength low alloy steels at a given value of tensile strength. Forming Limit Diagrams were determined for selected representative steels. Figure 3 is a plot of the relationship between the FLD_o point and gauge for VAN-QN. It is seen that the FLD_o points for the thinner gauge materials are somewhat higher than might be anticipated from the values characteristic for hot rolled materials. This suggests that the formability potential for cold rolled materials is at least as good as that observed for hot rolled materials realizing, of course, that FLD_o necessarily decreases as gauge is reduced.

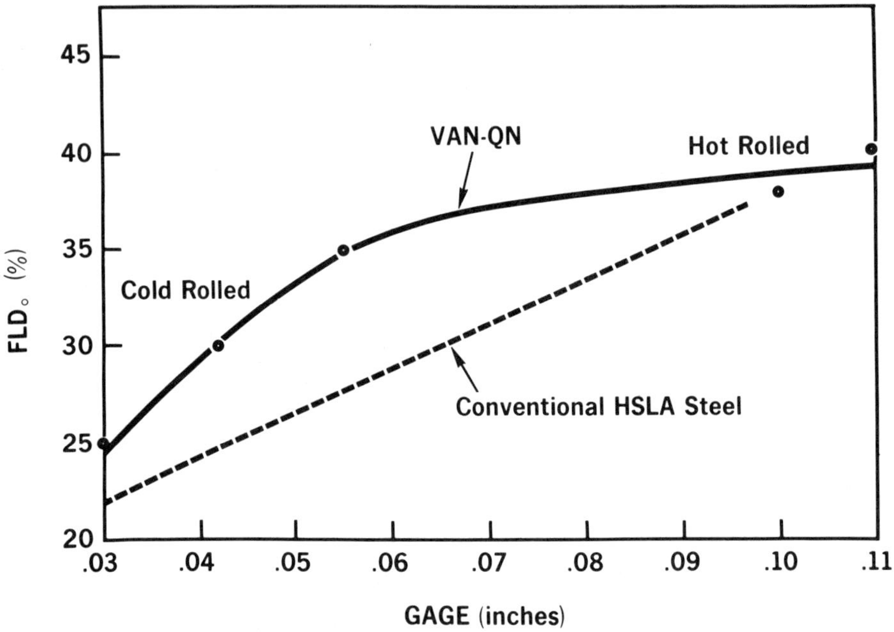

Figure 3. Superior formability of VAN-QN to SAE 950X is indicated especially in cold rolled gauges through this FLD_o analysis.

Products With New Strength-Ductility Combinations

Commerical utilization of dual-phase steel to date has largely centered upon a product with a specified minimum tensile strength of 90,000 psi in combination with a specified minimum total elongation of 27% (8). The data presented in Figure 1 will now be analyzed from the point of view of correspondence with existing theories and analyzed from the point of view of suggested new products. Figure 4 was constructed by drawing the best curve through the data of Figure 1 and utilization of certain information from the literature (9, 10). The experimental data were then converted to true stress-true strain values for ease of treatment. If we examine Figure 4 we see that in the range of 90,000 to 100,000 psi the ductility, as represented by n value, is significantly dependent upon stress level. However, as the stress level is increased the ductility tends to stabilize and become less dependent upon strength level. If we then extrapolate these

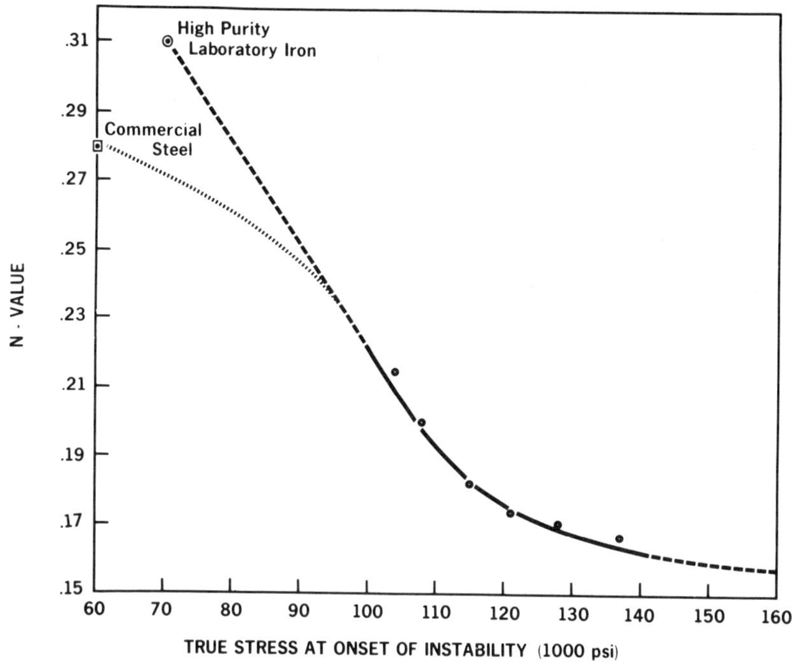

Figure 4. Relationship between true uniform strain and true stress at necking is developed using data from Figure 1 and the literature.

data to lower strength values going to a base value of the strength-ductility combination of pure ferrite, we see that a strong dependence of ductility on strength is indicated. These relationships are generally consistent with existing theories of dual-phase materials (11,12) and prior results (6), i.e. the tensile strength is a linear function of % MLB (martensite/lower bainite) to a first approximation with the ductility less than indicated by the law of mixtures.

The relationship of strength and ductility established in Figure 4 was then replotted in terms of the more typical engineering values for purposes of ease of discussion (Figure 5). Also entered in Figure 5 are the percentages of martensite or lower bainite consistent with the various strength-ductility combinations. Figure 5 will now be examined in terms of its implication of existing and potential commerical products.

Existing VAN-QN Product (90,000 minimum tensile strength, 27% minimum total elongation)

The applications of VAN-QN to date have by and large been difficult forming applications. It is seen that as the tensile strength increases above 90,000 psi a relatively significant decrease in ductility occurs. This clearly shows that for optimum formability the tensile strength must be controlled as close as possible to the 90,000 psi level. In practical terms the need for this specified minimum tensile strength must be placed within the context of good press shop performance and the realities of material

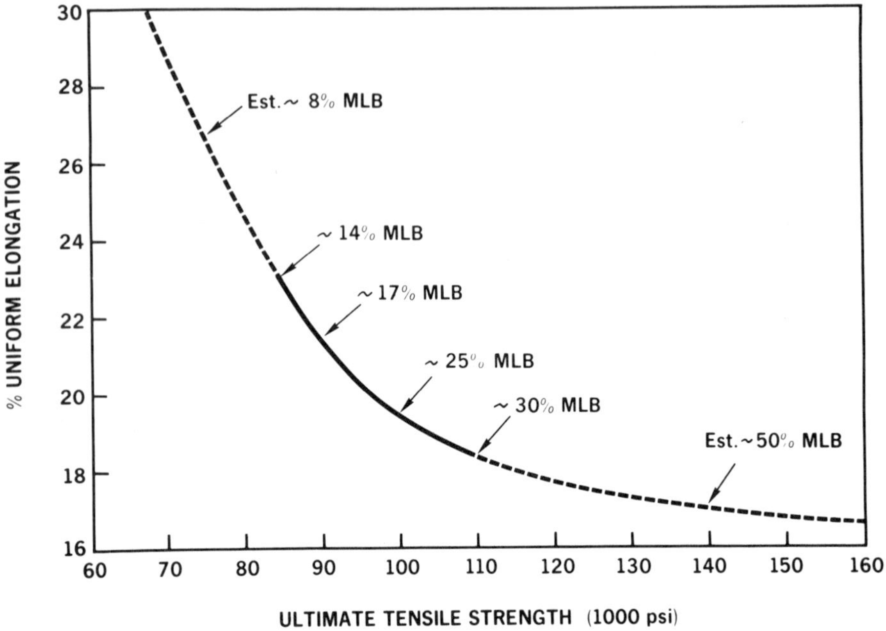

Figure 5. Re-plot of Figure 4 in terms of engineering stress and strain with % martensite/lower bainite indicated for various strength-ductility combinations.

production. A steel with a tensile strength of 105,000 psi, for example, may very well not satisfactory for many of the difficult forming applications presently encountered.

High Strength Product

However, when the tensile strength is in the range of 110,000 psi or greater significant increases in strength are indicated consistent with a relatively small decrease in ductility. This suggests that a significant potential exists for increased weight reduction using even higher strength material if the formability requirements are relaxed to accomodate such a material. Moreover, the percent martensite can vary over a relatively wide range with a relatively small change in strength and ductility. Clearly, the processing requirements are that much less difficult.

Low Strength Improved Formability Material

Materials with strengths in the range of 70 to 85,000 psi have significantly improved ductility. However, this strength range is obtained with materials containing a relatively small amount of martensite or lower bainite. Therefore, chemistry and process control requirements are quite stringent. However, this level of formability may, in fact, be necessary if the applications of dual-phase materials is extended into the sheet metal gauge range, for example, such parts as fenders, quarter panels, hoods, deck lids and the like. This type of material would be particularly valuable where relatively high levels of deformation are involved and the deformation is relatively uniform so that significant strengthening occurs throughout the formed part.

A Family of Dual-Phase Steels

Table III tabulates the tensile properties representative of conventional steels and low strength, medium strength and high strength dual-phase steels. It is seen that with a low strength dual-phase steel, identified in this table as VAN-QN (50), the strengthening potential in the formed part is consistent with a conventional 50,000 psi material, and the formability of a low strength commercial quality material is predicted. The substitutability of the existing VAN-QN product, VAN-QN (80), for a conventional SAE 980 steel has been demonstrated. Carrying this concept to the high strength range it is seen that if significant deformation is involved in the formed part, a dual-phase VAN-QN (110) has the basis for substitution for a 110,000 minimum yield strength material which is produced commercially by the quenching and tempering process. Clearly, in all cases the potential of dual-phase steels is dependent upon the forming characteristics of the part involved. Deformation is required to obtain the desired strengthening potential and obviously formability consistent with making the part is required. Materials in the low strength and high strength domain have not been made on a production basis to a degree sufficient to fully substantiate the potential indicated. However, materials produced in our laboratory and examination of the literature suggests that these materials are commercially viable. Evaluation of materials of these types on production parts is clearly the next step to determine their potential.

TABLE III

COMPARISON OF TENSILE PROPERTIES OF CONVENTIONAL STEELS WITH DUAL-PHASE STEELS AT DIFFERENT STRENGTH LEVELS

Steel	Yield Strength (1000 psi)	Ultimate Tensile Strength (1000 psi)	% Uniform Elongation	% Total Elongation	Comments
Commercial Quality	25/40	35/50	22/27	32/38	Based upon cold rolled steel (~.03" thick).
Drawing Quality	25/35	35/45	25/30	35/40	Based upon cold rolled steel (~.03" thick).
SAE 950	50/60	65/75	15/18	20/25	Based upon cold rolled steel (~.03" thick).
VAN-QN (50)	35/50	70/80	25/28	30/35	Based upon cold rolled steel (~.03" thick).
SAE 980	80/90	95/105	11/13	19/22	Based upon hot rolled steel (~.1" thick).
VAN-QN (80)	50/65	90/100	20/23	27/30	Based upon hot rolled steel (~.1" thick).
Q&T 110	110/120	120/130	10/12	16/19	Based upon hot rolled steel (~.1" thick).
VAN-QN (110)	55/70	115/130	15/18	19/22	Based upon hot rolled steel (~.1" thick).

Summary

This paper has updated previous strength/formability characterization of VAN-QN. It has been shown that materials in cold rolled gauges as thin as .022" and galvanized steels can be produced with properties essentially identical with those of previously described hot rolled materials. Moreover, in addition to the specific product in commercial use today it has been shown that lower strength and higher strength variations of this product have significant potential for specific applications. The lower strength variety would have its particular value where formability requirements were most demanding. The higher strength product, though somewhat limited in formability, has significant weight reduction possibilities in view of its very high strength. Clearly, utilization of these materials will be dependent upon their cost effectiveness. Extension into the lower and higher strength domains is based upon essentially the same low alloy chemistry and standard processing procedures. Therefore, these products should be relatively economical in comparison with alternative materials whether they be steel or non-ferrous materials for the applications suggested.

References

1. Cairns, R. L. and Charles, J. A., JISI, 1967, Vol. 205, p. 1051.

2. Grange, R. A., 2nd International Conference on Strength of Metals and Alloys, 1970, p. 861.

3. Araki, K., et al., "Development and Features of Continuous Annealed High Strength Cold Rolled Sheet Steels in Japan", IDDRG Proceedings, 1976, p. 39.

4. Rashid, M. S., "GM 980X - A Unique High Strength Sheet Steel with Superior Formability, SAE, 1976, 760206.

5. Davies, R. G., "The Deformation Behaviour of a Vanadium-Strengthened Dual Phase Steel", Met. Trans. A., Vol. 9A, 1978, p. 41.

6. Bucher, J. H. and Hamburg, E. G., "High Strength Formable Sheet Steel", SAE, 1977, 770164.

7. Rashid, M. S., SAE Preprint, 770211, February 1977.

8. General Motors Engineering Standards, High Strength Low Alloy Sheet Steel, GM 6187-M.

9. Bucher, J. H. and Goodenow, R. H., "Yielding and Flow Characteristics of Low-C Steel Between Ambient and Liquid Nitrogen Temperatures", Journal of Basic Engineering, p. 603, December 1969.

10. Morrison, W. B., "The Effect of Grain Size on the Stress-Strain Relationship in Low-C Steel", Trans. ASM, Vol. 59, 1966, p. 824.

11. Mileiko, S. T., Journal of Material Science, 1969, Vol. 4, p. 974.

12. Garmong, G. and Thompson, R. B., TAIME, 1973, Vol. 4, p. 863.

INTERCRITICALLY ANNEALED DUAL-PHASE STEELS

FOR AUTOMOTIVE APPLICATIONS

J. W. Morrow,* G. Tither** and R. M. Buck***

Climax Molybdenum Company of Michigan
A Subsidiary of AMAX Inc.
Ann Arbor, Michigan

Abstract

Reheating steel strip into the γ/α region for short periods of time followed by air cooling can, in certain steel compositions, result in transformation to a polygonal ferrite matrix with dispersed martensite and/or bainite islands. These steels, commonly referred to as dual-phase steels, possess excellent formability properties and are capable of work hardening from a low initial yield strength [<345 MPa (50 ksi)] to a strength of up to about 550 MPa (80 ksi) after only 2% strain. Tensile strength is typically about 655 MPa (95 ksi) and elongation values up to 31% have been recorded.

This paper reports on the development of a family of molybdenum-containing steels which respond favorably to intercritical annealing for very short periods of time at temperature. The effects of processing variables, including annealing time (15 secs to 180 secs), annealing temperature [790 to 830°C (1450 to 1525°F)], and cooling rate following annealing, on mechanical properties and microstructure are discussed in detail. These low carbon Mn-Si-Mo steels offer considerable potential as highly formable, HSLA steels for many automotive parts.

*Research Associate; **Manager, High Strength Steel Development; and ***Manager, Automotive Market Development.

Introduction

Renewed interest in high strength, low alloy (HSLA) steels for automotive applications has been stimulated by the need for increased fuel economy through higher strength and lighter gauges, thereby decreasing the overall weight of the vehicle. However, to effectively utilize HSLA steels, their formability must be comparable to that of plain carbon steel. This challenge has led to the recent development[1-6] of dual-phase steels. These steels exhibit continuous yielding behavior, a low initial yield strength, and high ductility which results in formability similar to that of low strength steels, and they have a high work hardening rate which results in high strength in the formed part.

The continuous yielding behavior, rapid work hardening rate, and improved elongation of the dual-phase steels, compared with the discontinuous yielding in typical HSLA automotive steels, is illustrated in Figure 1. The term "dual-phase" relates to the microstructure of about 80-90% polygonal ferrite with 10-20% dispersed martensite islands containing austenite (M-A constituent). The stress-strain behavior of the dual-phase steels results from the residual stresses formed in the ferrite due to the volume and shape change during the martensitic transformation. These residual stresses, plus mobile dislocations allow early and uniform yielding of the strip. A typical dual-phase microstructure, as observed in the scanning electron microscope (SEM), is shown in Figure 2.

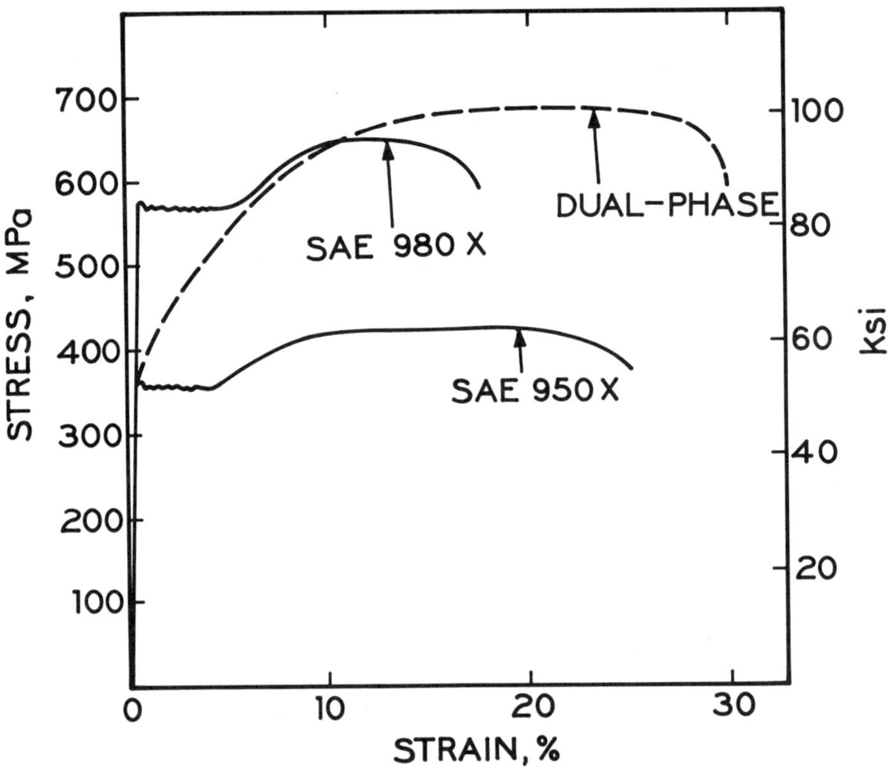

Figure 1 Typical Stress-Strain Curves for High-Strength Automotive Strip

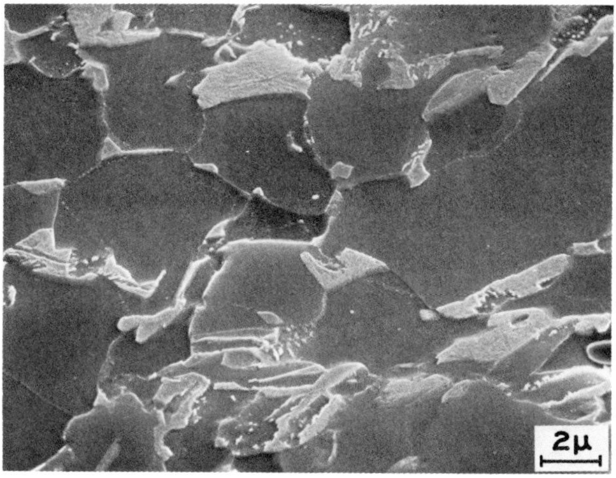

Figure 2 Scanning Electron Micrograph of a Typical Dual-Phase Steel. Picral-Nital Etch. Original Magnification 5000X.

In effect, these steels behave like a low strength steel of 275 to 345 MPa (40 to 50 ksi) yield strength during initial forming operations, increasing in strength during each forming operation where strain is introduced. The final formed part, therefore, typically exhibits a yield strength of 550 MPa (80 ksi).

The purpose of this paper is to report recent results pertaining to intercritically annealed, Mn-Si-Mo dual-phase steels. Laboratory studies conducted on several molybdenum-containing steels showed that these materials respond very well to continuous annealing processes to produce high strength strip for automotive applications. Compositions examined included 0.08-0.11% C, 1.27-1.49% Mn, 0.23-0.86% Si, 0.0-0.32% Mo, 0.0-0.04% Nb, and 0.0-0.07% V.

Experimental Procedures

Melting and Rolling

Laboratory steels were prepared as 34-kg (75-lb) heats melted and cast under an argon atmosphere to control nitrogen content. The chemical compositions of the twelve steels studied are given in Table I.

From each steel, 25-mm (1-inch) thick by 125-mm (5-inch) wide by 50-mm (2-inch) long rolling blanks were obtained and rolled to 2.5-mm (0.100-inch) thick strips in six passes using a hot-strip mill simulation. This consisted of a slab reheat temperature of 1260°C (2300°F), a finish rolling temperature of 870°C (1600°F), accelerated cooling to a coiling temperature of 595°C (1100°F), then slow cooling at a rate of 28 C°/hr (50 F°/hr) to below 315°C (600°F).

Intercritical Annealing

From each hot-rolled strip, full-thickness, longitudinal tensile specimens with a 13-mm (0.500-inch) wide by 57-mm (2.25-inch) long reduced section

TABLE I

Compositions of Experimental Steels

Steel No.	Element, wt. %									
	C	Mn	Si	Mo	Nb	V	P	S	N	Al
1	0.11	1.44	0.54	--	0.04	--	0.008	0.006	0.005	0.05
2	0.11	1.45	0.55	0.21	0.04	--	0.009	0.006	0.005	0.05
3	0.11	1.44	0.54	0.32	0.04	--	0.008	0.006	0.004	0.05
4	0.12	1.43	0.23	0.22	0.04	--	0.008	0.005	0.005	0.06
5	0.11	1.45	0.86	0.21	0.04	--	0.008	0.005	0.004	0.05
6	0.11	1.43	0.55	0.10	--	0.07	0.009	0.006	0.008	0.05
7	0.11	1.27	0.57	0.15	--	--	0.009	0.005	0.008	0.06
8	0.11	1.49	0.56	0.15	--	--	0.010	0.005	0.008	0.06
9	0.08	1.43	0.54	--	--	--	0.006	0.006	0.005	0.05
10	0.08	1.44	0.56	0.10	--	--	0.006	0.006	0.005	0.04
11	0.11	1.46	0.55	--	--	--	0.006	0.006	0.005	0.06
12	0.11	1.46	0.56	0.10	--	--	0.006	0.006	0.005	0.05

were prepared. Specimens were intercritically heat treated at 790°C (1450°F) for times of 15 seconds up to 20 minutes followed by either air cooling at 5 C°/sec (9 F°/sec) or accelerated cooling at 28 C°/sec (50 F°/sec). Specimens from Steel 8 were also intercritically annealed at 800, 815, and 830°C (1475, 1500, and 1525°F) for times of 15 seconds or 1 minute and air cooled.

Tensile Testing

Duplicate room temperature tensile tests were conducted for each annealing condition using strain rates of 18% and 120% per hour in the elastic and plastic ranges, respectively. Total elongation values were determined over a 50-mm (2-inch) gauge length by reassembling the broken specimen and measuring the separation with vernier calipers.

Metallography

Specimens for optical and scanning electron microscopy were sectioned from tensile specimens, mechanically polished, and etched. Specimens were sectioned so that the surface to be examined was parallel to the direction of rolling and perpendicular to the plane of rolling. Preparation for optical microscopy involved etching in a 2% nital solution, while specimens for the SEM were etched in a solution of 4% picric acid-1% nitric acid in ethanol.

A gold-palladium film approximately 10 to 20 nm (100 to 200 Å) thick was vapor deposited on the surface of the SEM samples to prevent charge build-up.

Results

Microalloyed Steels

The first six steels (Steels 1-6) listed in Table I represent the initial compositions examined in the laboratory. All contained either niobium or vanadium.

Figure 3 shows the effect of intercritical annealing on the stress-strain behavior of Steel 1 (molybdenum-free). In the as-rolled condition this steel exhibited typical, discontinuous yielding with a 0.2% offset yield strength of about 500 MPa (75 ksi). After intercritically annealing for one minute at 790°C (1450°F) the yield strength decreased only slightly, and the stress-strain behavior was still discontinuous. In comparison, the same steel containing 0.21% Mo (Steel 2) exhibited a continuous stress-strain behavior with a relatively low 0.2% offset yield strength of about 345 MPa (50 ksi) after intercritically annealing for one minute at 790°C (1450°F), as shown in Figure 4.

The effect of molybdenum content on the tensile properties after intercritically annealing for one minute at 790°C (1450 °F) is further illustrated in Figure 5. The addition of 0.21% Mo reduced the 0.2% offset yield strength while increasing both the 2% offset yield strength and the tensile strength. Increasing the molybdenum content to 0.32% further increased the 2% offset yield strength and tensile strength without significantly affecting the 0.2% offset yield strength.

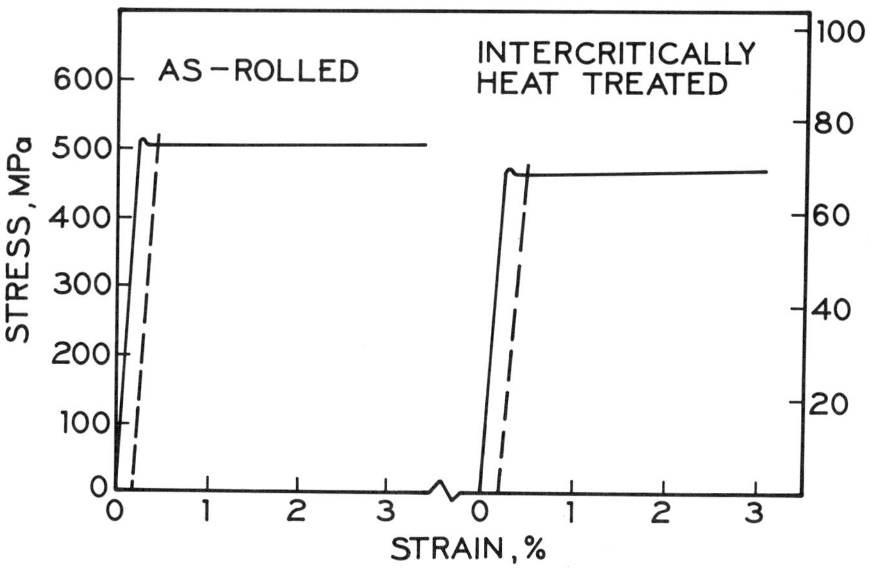

Figure 3 Stress-Strain Curves for Steel 1 (Molybdenum-Free) in the As-Rolled Condition and Intercritically Annealed for One Minute at 790°C (1450°F)

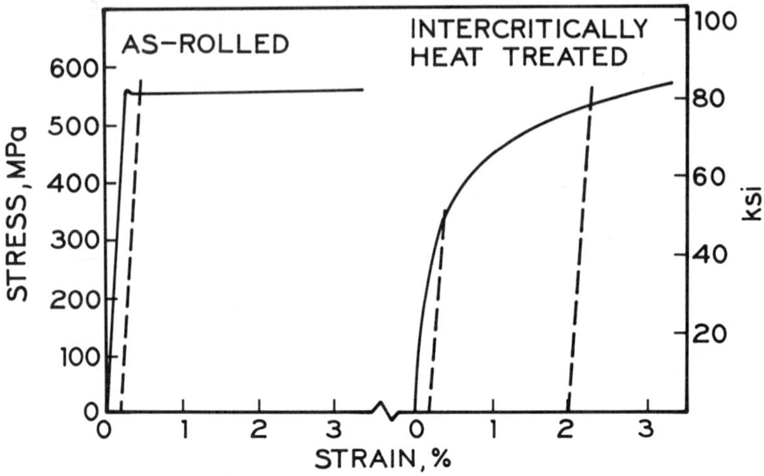

Figure 4 Stress-Strain Curves for Steel 2 (0.21% Mo) in the As-Rolled Condition and Intercritically Annealed for One Minute at 790°C (1450°F).

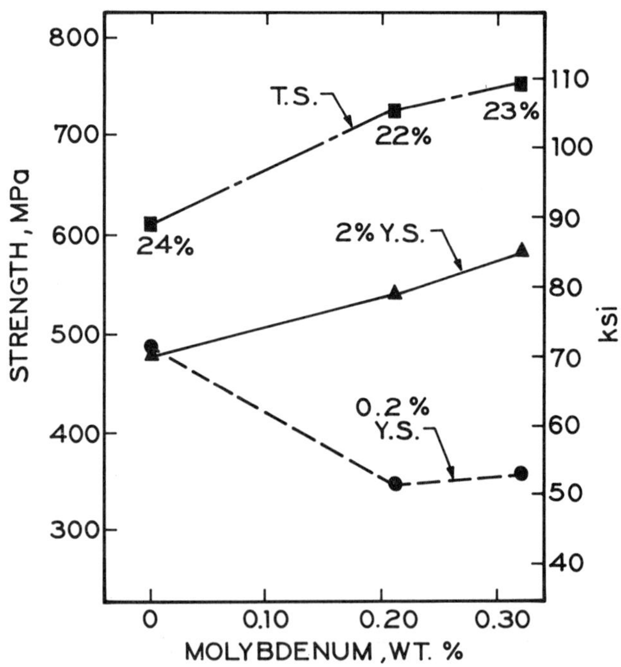

Figure 5 Effect of Molybdenum Content on the Tensile Properties of 2.5-mm (0.100-inch) Strip Intercritically Annealed for 1 Minute at 790°C (1450°F) and Air Cooled. Base Composition: 0.11% C - 1.45% Mn - 0.55% Si - 0.04% Nb. Numbers below Tensile Strength Data Points Indicate Total Elongation

Scanning electron micrographs of the molybdenum-free and the 0.21% Mo steels are shown in Figure 6. The microstructure of the molybdenum-free steel consisted of fine-grained polygonal ferrite with dispersed cementite particles and areas of pearlite or upper bainite. In contrast, the 0.21% Mo steel exhibited islands of martensite in place of the carbide-rich areas in the molybdenum-free steel.

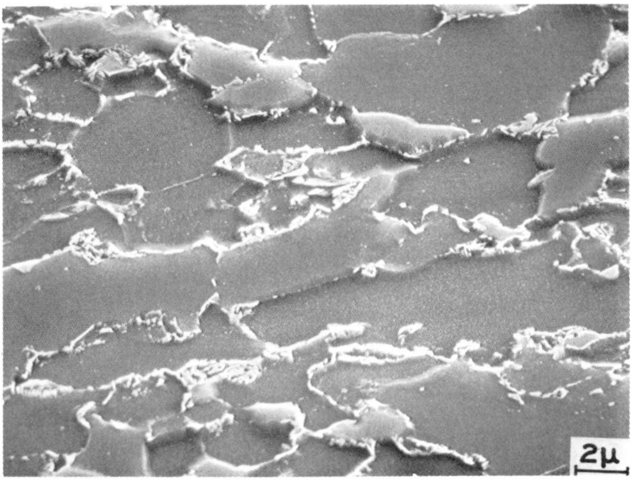

(a) 0% Mo (Steel 1)

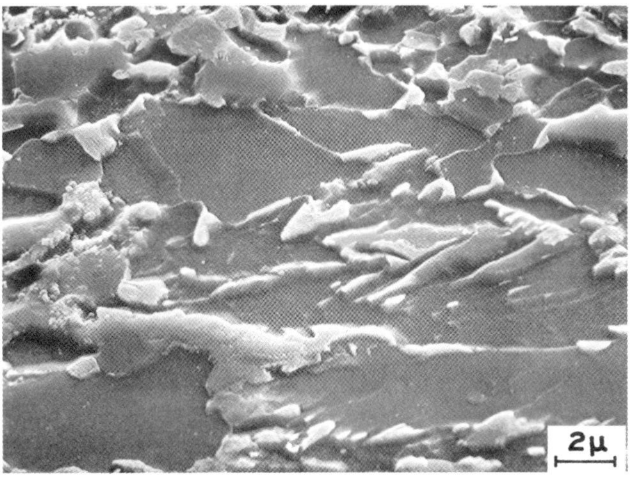

(b) 0.21% Mo (Steel 2)

Figure 6 Scanning Electron Micrographs of Steels 1 and 2 Intercritically Annealed at 790°C (1450°F) for 1 Minute and Air Cooled. Picral-Nital Etch. Original Magnification 5000X.

The effect of silicon content on the tensile properties of a 0.11% C - 1.45% Mn - 0.21% Mo - 0.04% Nb steel after intercritically annealing for one minute at 790°C (1450°F) is shown in Figure 7. At all three levels of silicon (0.23, 0.55 and 0.86%) the 0.2% offset yield strength was about 345 MPa (50 ksi), but increasing silicon content raised both the 2% offset yield stress and the tensile strength by about 16 MPa (2.3 ksi) per 0.1% Si.

All of the data previously discussed were for intercritical annealing times of one minute. However, the data for annealing times from 15 seconds to 20 minutes (shown in Figure 8 for Steel 2) indicate that the Mo-Nb steels investigated were insensitive to annealing time. Over this range of time the 0.2% offset yield strength, the 2% offset yield strength, and the tensile strength varied by only about ±25 MPa (3.5 ksi) from their respective average values. The Mo-V steel (Steel 6) was also relatively insensitive to annealing time, although there seemed to be a tendency toward higher strengths with longer annealing times, as shown in Figure 9.

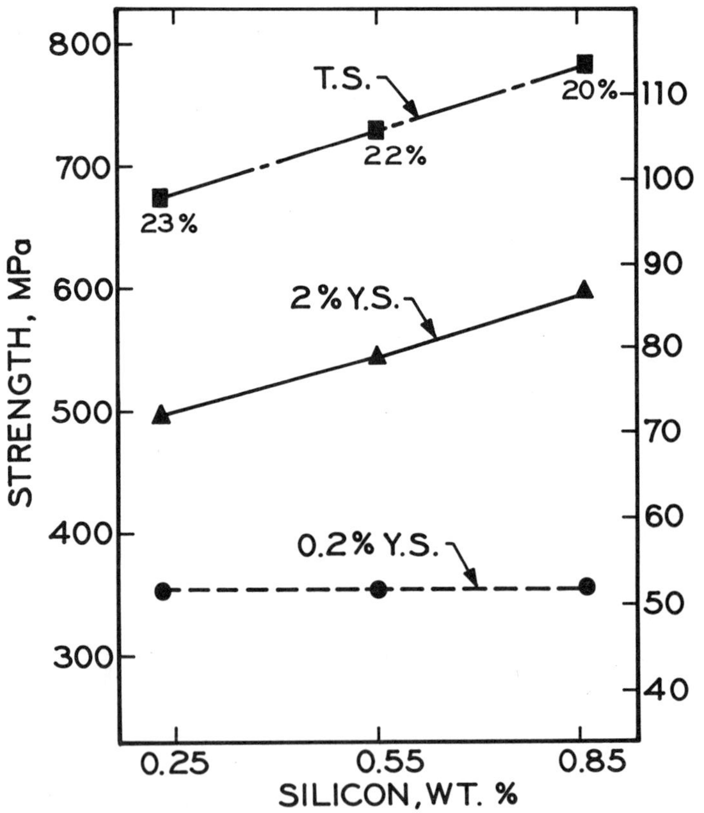

Figure 7 Effect of Silicon Content on the Tensile Properties of 2.5-mm (0.100-inch) Strip Intercritically Annealed for 1 Minute at 790°C (1450°F) and Air Cooled. Base Composition: 0.11% C - 1.45% Mn - 0.21% Mo - 0.04% Nb. Numbers below Tensile Strength Data Points Indicate Total Elongation.

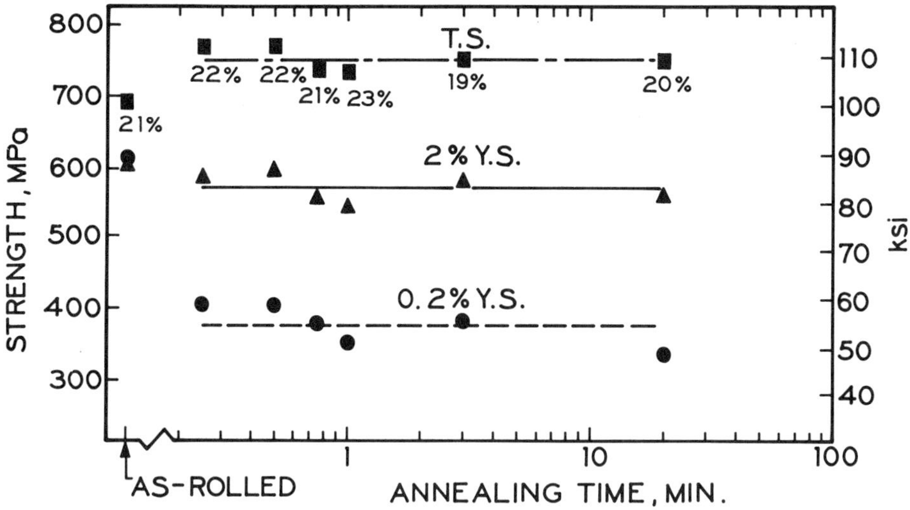

Figure 8 Effect of Annealing Time on the Tensile Properties of Steel 2 Intercritically Annealed at 790°C (1450°F) and Air Cooled. Numbers below Tensile Strength Data Points Indicate Total Elongation.

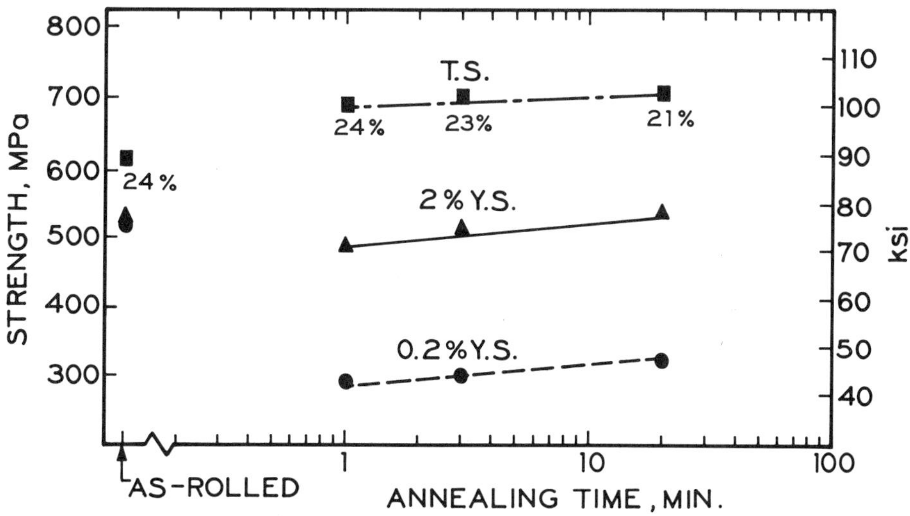

Figure 9 Effect of Annealing Time on the Tensile Properties of Steel 6 Intercritically Annealed at 790°C (1450°F) and Air Cooled. Numbers below Tensile Strength Data Points Indicate Total Elongation.

The work on microalloyed dual-phase steels indicated that molybdenum played an important role in achieving the desired microstructure, stress-strain behavior, and strength levels in samples intercritically annealed at 790°C (1450°F) and air cooled. Molybdenum increased the hardenability of the high carbon austenite islands formed during intercritical annealing and caused transformation to martensite rather than to pearlite or bainite. The Mo-Nb steels were found to be insensitive to intercritical annealing time since all of the molybdenum is in solid solution and the only time necessary is for the solution of cementite. The slight sensitivity of the Mo-V steel to annealing time can be attributed to the time required for vanadium to dissolve in the austenite where it can act as a hardenability agent.

Good strength levels were obtained in all of the molybdenum-containing microalloyed steels; however, total elongation values in the range of 19 to 24% were achieved in the Mo-Nb and Mo-V steels. These were lower than the 27% minimum that had been suggested[3] for automotive applicatons. Since the problem of obtaining a dual-phase microstructure was primarily a question of hardenability, it was believed that the microalloying elements (particularly niobium) played a minor (if not detrimental) role in developing the desired tensile properties. With this in mind, the second stage in this development of dual-phase steels was concerned with microalloy-free steels.

Mn-Si-Mo Steels

The second six steels (Steels 7-12) listed in Table 1 represent the microalloy-free steels examined. All were prepared as 2.5-mm (0.100-inch) strip as previously described.

The effect of manganese content on the tensile properties of a steel having a base composition of 0.11% C - 0.56% Si - 0.15% Mo is shown in Table II. Both steels were intercritically annealed for one minute at 790°C (1450°F) and air cooled. Offset yield strengths (0.2%) of about 290 MPa (42 ksi) were obtained for both steels, with the 2% offset yield strength and tensile strengths being about 50 MPa (7 ksi) higher for the higher manganese steel. The 1.27% Mn steel achieved a tensile strength of 620 MPa (90 ksi), and elongation values of 28 and 29% were obtained in the two steels.

TABLE II

Effect of Manganese Content on the Tensile Properties of 2.5-mm (0.100-inch) Strip Intercritically Annealed at 790°C (1450°F) for 1 Minute and Air Cooled

Steel No.	Mn Content, %	0.2% Y.S., MPa (ksi)	2% Y.S. MPa (ksi)	T.S., MPa (ksi)	Elongation, %
7	1.27	291 (42.2)	431 (62.5)	620 (90.0)	29
8	1.49	288 (41.8)	487 (70.6)	671 (97.3)	28

Base Composition: 0.11% C - 0.56% Si - 0.15% Mo

The effect of intercritical annealing temperature on the tensile properties of Steel 8 (0.11% C - 1.49% Mn - 0.56% Si - 0.15% Mo) is illustrated in Figure 10. The overall strength level of this steel decreased slightly as the intercritical annealing temperature was increased; and differences in strength due to annealing time were very small. The effect of temperature on total elongation was somewhat inconclusive; however, after intercritically annealing this steel for 1 minute at 830°C (1525°F), the total elongation was 29%, and in general, it appears that improved elongation values may result from the higher intercritical annealing temperatures in this steel.

Microstructures resulting from the four intercritical annealing temperatures are shown in Figure 11. All were similar, with the only difference being a slight coarsening of the ferrite grain size from ASTM 12 at 790°C (1450°F) to ASTM 10.5 at the highest intercritical annealing temperature of 830°C (1525°F). The volume percent M-A constituent remained relatively constant at about 20% for all four temperatures. The coarsening of the ferrite grains correlates well with the slight decrease in tensile properties and the small increase in total elongation.

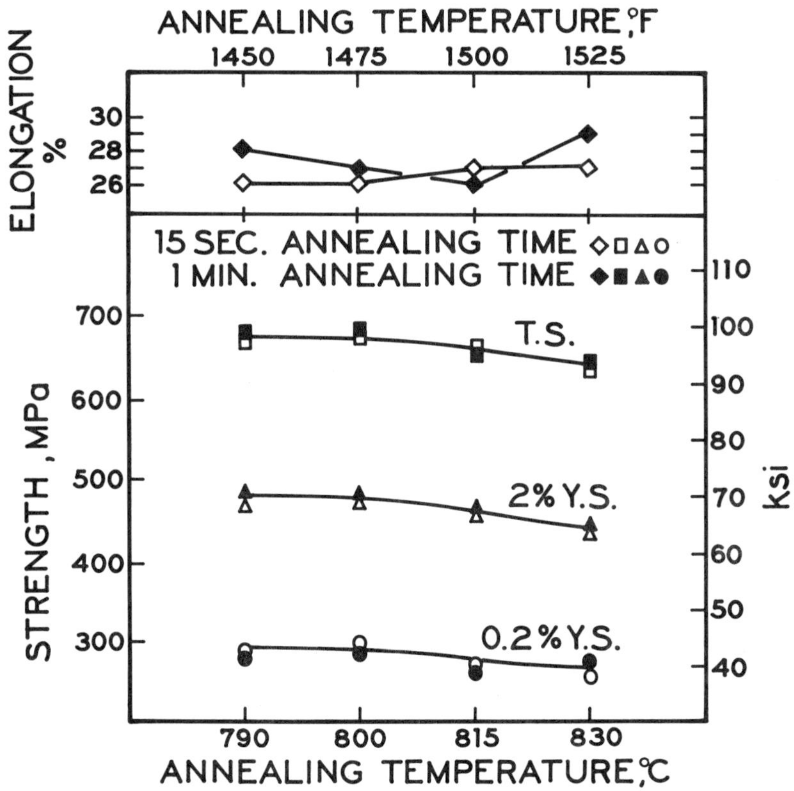

Figure 10 Effect of Intercritical Annealing Temperature on the Tensile Properties of Steel 8. Samples were Air Cooled at 5 C°/sec (9 F°/sec) after Intercritical Annealing.

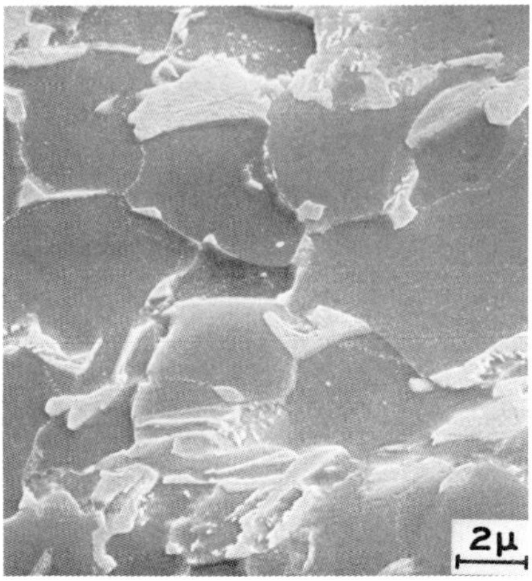

(a) 790°C (1450°F)

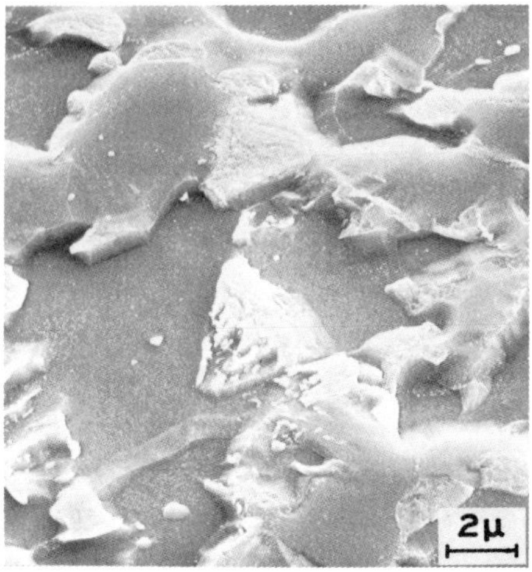

(b) 800°C (1475°F)

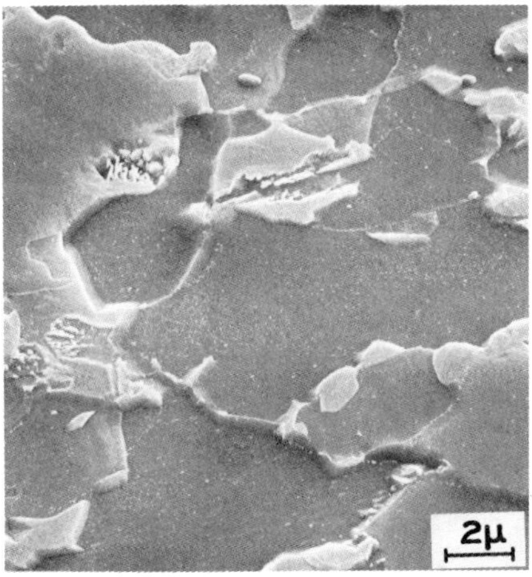

(c) 815°C (1500°F)

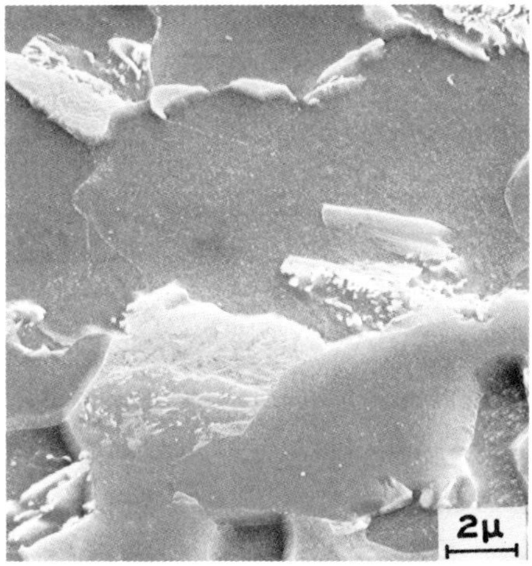

(d) 830°C (1525°F)

Figure 11 Scanning Electron Micrographs of Steel 8 Intercritically Annealed for 1 Minute and Air Cooled. Picral-Nital Etch. Original Magnification 5000X.

The effect of carbon content, molybdenum content and cooling rate following intercritical annealing on the tensile properties of Steels 9-12 is illustrated in Figure 12. In the air cooled condition [cooling rate of 5 C°/sec (9 F°/sec)] only the 0.11% C steel containing molybdenum (Steel 12) produced the desired dual-phase properties, with a tensile strength of 660 MPa (96 ksi) and total elongation of 31%. The two molybdenum-free steels (Steels 9 and 11) plus the 0.08% C - 0.10% Mo steel (Steel 10) failed to develop acceptable properties after air cooling. After accelerated cooling at 28 C°/sec (50 F°/sec) following intercritical annealing, only the 0.08% C, molybdenum-free

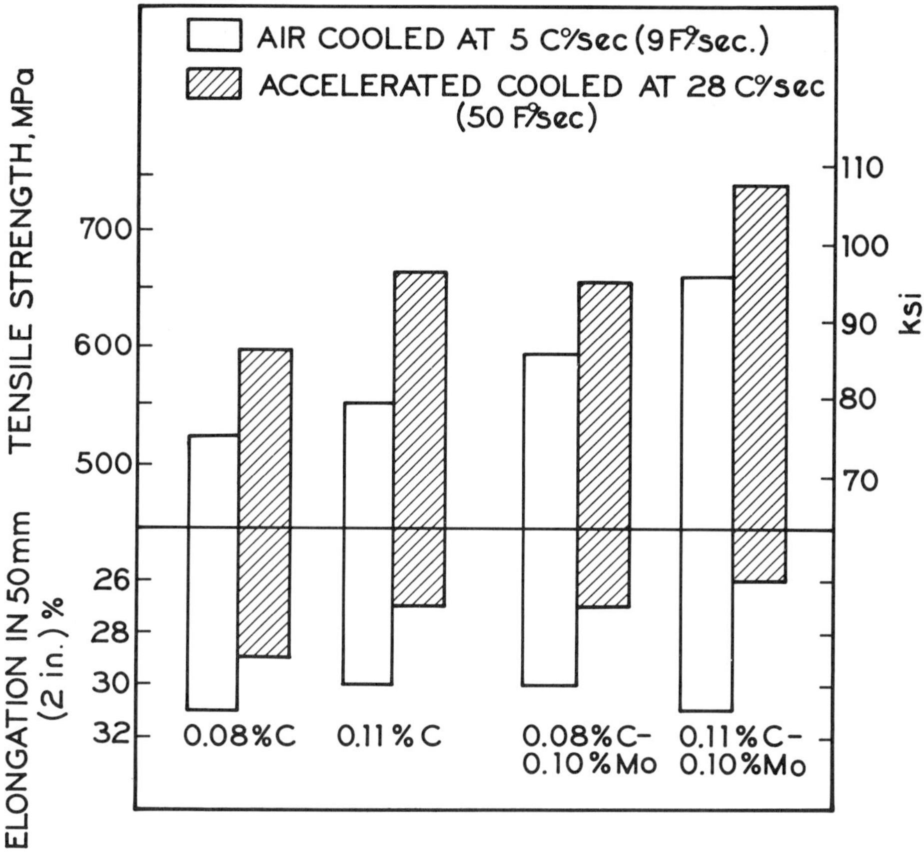

Figure 12 Effect of Carbon Content, Molybdenum Content, and Cooling Rate on the Tensile Strength and Elongation of 2.5-mm (0.100-inch) Strips Intercritically Annealed for 1 Minute at 790°C (1450°F).

steel (Steel 9) did not exhibit the desired dual-phase properties. The remaining three steels exhibited tensile strengths in the range 655 to 750 MPa (95 to 109 ksi) with total elongations of 26 to 31%. The best combinations of tensile strength and elongation were obtained in the molybdenum-containing steels which were air cooled after intercritical annealing. In the 0.08% C steel, an accelerated cooling rate plus 0.10% Mo was necessary to achieve the desired properties. In the 0.08% C steels, the additon of 0.10% Mo increased the tensile strength by an average of about 65 MPa (9.5 ksi), while in the 0.11% C steels, the tensile strength increased by an average of about 95 MPa (14 ksi).

The effect of adding 0.10% Mo on the resultant microstructure of the 0.11% C steels is illustrated in Figure 13. In the molybdenum-free steel a structure consisting of ferrite, pearlite, and grain boundary cementite was obtained on air cooling from the intercritical temperature. In contrast, the addition of 0.10% Mo increased the hardenability of the austenite formed during intercritical annealing and promoted the formation of the desired low temperature transformation products on cooling. A similar effect occurred in the lower carbon steels under accelerated cooling conditions. In all six steels, grain sizes of intercritically annealed specimens were in the range of ASTM 12-13.

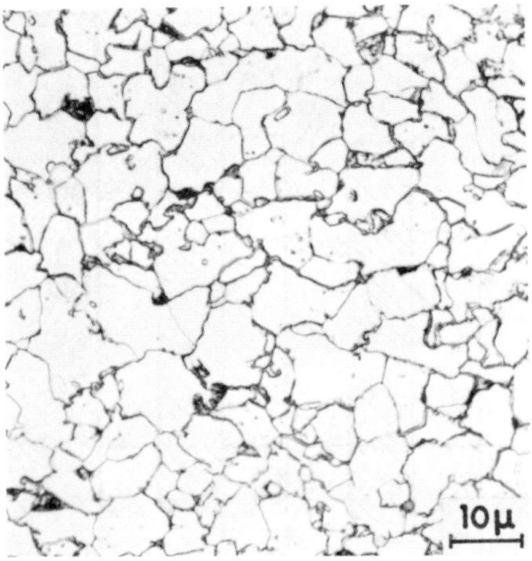

(a) 0% Mo - Optical (Steel 11)

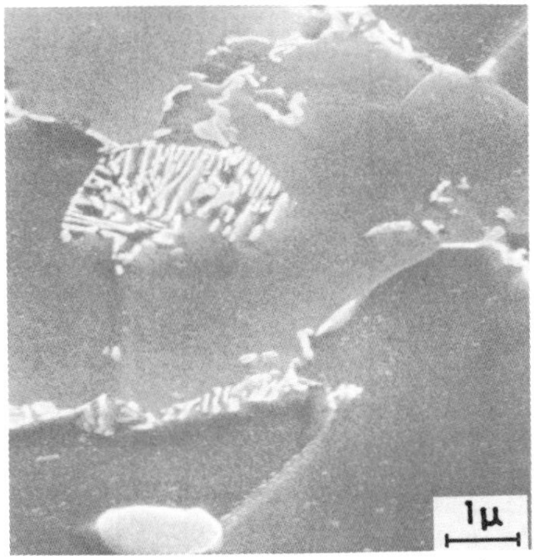

(b) 0% Mo - SEM (Steel 11)

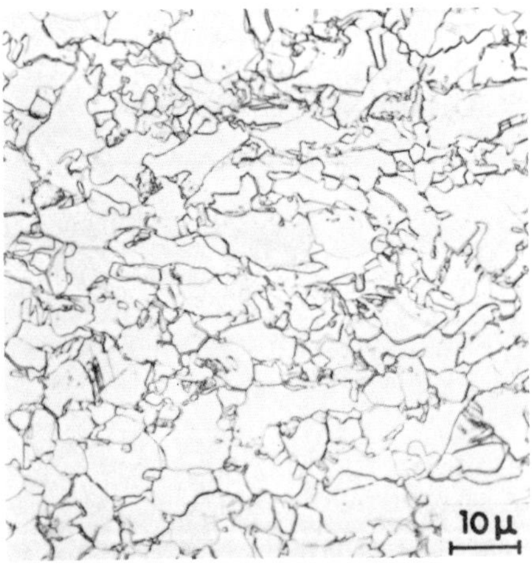

(c) 0.10% Mo - Optical (Steel 12)

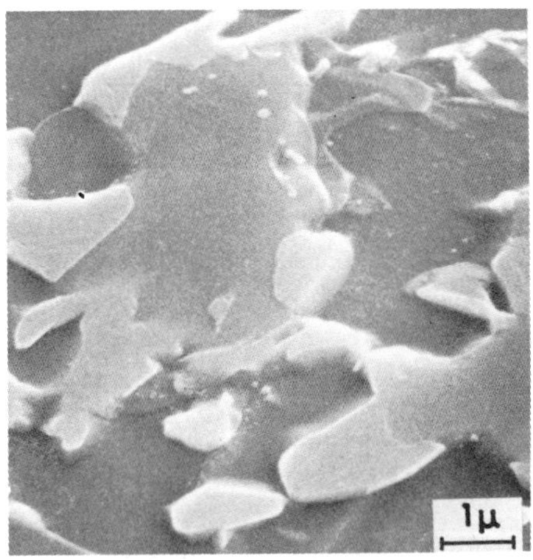

(d) 0.10% Mo - SEM (Steel 12)

Figure 13 Optical and Scanning Electron Micrographs Showing the Effect of 0.10% Mo on the Microstructure of Steels Intercritically Annealed at 790°C (1450°F) for 1 Minute and Air Cooled. 2% Nital Etch for (a) and (c), Picral-Nital Etch for (b) and (d). Original Magnifications 1000X for (a) and (c), 10,000X for (b) and (d).

Summary and Conclusions

The demand for high strength, formable steel strip for use by the automotive industry is certain to increase as the requirements for improved gas mileage become more stringent. The weight reductions which can be achieved with the use of the dual-phase steels being developed offer a cost-effective method for decreased weight and increased fuel economy. The current laboratory program has shown that molybdenum-containing steels produced as intercritically annealed strip can meet the requirements of low initial yield strength, high tensile strength, and good ductility for applicatons such as bumpers, wheel rims, spiders, control arms, and other automotive components which demand a combination of high strength and good formability.

Results on a series of microalloyed dual-phase steels showed that the Mo-Nb steels were insensitive to intercritical annealing time in the range of 15 seconds to 20 minutes. The tensile properties of a Mo-V composition had a tendency to be somewhat time dependent, and this was attributed to the time necessary to dissolve V(C,N) at the intercritical temperature. Increasing silicon content raised both the 2% yield strength and the tensile strength without affecting the 0.2% offset yield strength. Total elongation values of the Nb or V microalloyed steels were lower than the 27% minimum desired. This is believed to be due to the deleterious effect of niobium or vanadium precipitates on ductility.

Results on Mn-Si-Mo steels indicated that the addition of as little as 0.10% Mo helped promote ferrite-martensite structures in steels air cooled from the intercritical annealing temperature. It was shown that intercritical annealing time had little effect on properties, and that good properties could be obtained after only 15 seconds at temperature. Furthermore, within the range of 790 to 830°C (1450 to 1525°F), little variation in properties occurred. Accelerated cooling was shown to enhance the properties of those steels which exhibited marginal or unacceptable properties when air cooled. The current laboratory program on the Mn-Si-Mo steels has recently been adapted to commercial practice by McLouth Steel Corporation[6] with excellent results.

The molybdenum-containing steels consistently exhibited acceptable properties when cooling conditions and alloy content were properly adjusted. From the current work, a recommended commercial melting range would be 0.08/0.11% C - 1.25/1.50% Mn - 0.45/0.60% Si - 0.10/0.15% Mo. Cooling rates following intercritical annealing in the range of 790 to 830°C (1450 to 1525°F) would be adjusted from air cooling for the richest compositions to approximately 28 C°/sec (50 F°/sec) for the leaner compositions, especially the lower carbon steels. Typical properties to be expected would be in the range of 620 to 690 MPa (90 to 100 ksi) tensile strength and 26 to 31% elongation.

References

1. Hayami, S. and Furakawa, T., Proceedings of Micro Alloying '75, Washington, D. C., October 1-3, 1975, p. 311.

2. Rashid, M. S., "GM 980X-A Unique High Strength Sheet Steel with Superior Formability," SAE, Detroit, MI, Paper No. 760206, 1976.

3. Rashid, M. S., "GM 980X-Potential Applications and Review," SAE, Detroit, MI, Paper No. 770211, 1977.

4. Bucher, J. H. and Hamburg, E. G., "High Strength Formable Sheet Steel," SAE, Detroit, MI, Paper No. 770164, 1977.

5. Morrow, J. W. and Tither, G., "Molybdenum in Intercritically Annealed Dual-Phase Steel Strip," Paper submitted to Journal of Metals.

6. McFarland, W., McLouth Steel Corporation, Detroit, MI, private communication.

RECENT DEVELOPMENTS IN FORMABLE HOT- AND COLD-ROLLED HSLA INCLUDING DUAL-PHASE SHEET STEELS

S. Hayami,* T. Furukawa,* H. Gondoh** and H. Takechi**

* Fundamental Research Laboratories, Nippon Steel Corporation, Kawasaki, Kanagawa, Japan

** Technical Research Center, Kimitsu Works, Nippon Steel Corporation, Kimitsu, Chiba, Japan

Abstract

A production technique for high-strength steels with superior formability has been extensively developed at Nippon Steel Corporation. In addition to conventional types of HSLA steels, hot- and cold-rolled dual-phase sheet steels covering 500 to 800 MPa in tensile strength are currently produced at mills with heat-treating lines, using plain carbon-silicon-manganese steels of low carbon content ($C \leq 0.1\%$). These steels exhibit yield to tensile ratios of 0.5 to 0.6, and n values of ≥ 0.2. Mild cooling rates (5 - 20 degrees C/second) after dual-phase treatment are preferred to drastic quenching from the viewpoint of elongation, presumably due to cleaner ferrite with respect to carbon in solution. Contrary to vanadium-containing steels (e.g., VAN-QN), silicon-manganese steels increase their tensile strength while keeping their originally good elongation, after dual-phase treatment. As a variation in the dual-phase category, a 1000 MPa grade in tensile strength has also been developed, which consists of un-recrystallized ferrite and martensite.

1. Introduction

Major drawbacks in the cold forming of HSLA steels are considered to be springback and poor stretchability.[1] Increased attention is directed towards the dual-phase steels as an attractive answer to the problem, because of their lower yield strength, higher work-hardening rate and larger elongation, compared to conventional HSLA steels.[2~5]

Developments in the production technique of dual-phase steels from both cold-rolled and hot-rolled gages are notable at Nippon Steel. Most of the dual-phase steels are based on plain carbon-silicon-manganese steels with no additions of carbo-nitride forming elements.

Although Nippon Steel produces other types of HSLA steel, such as solid-solution hardening steels having large $\bar{r}$ values, and conventional precipitation hardening steels, emphasis is laid on the production technology of dual-phase steels in this paper.

2. Cold-Rolled HSLA Steels

Table I shows some typical mechanical properties of cold-rolled HSLA steels. This includes the solid-solution hardening type characterized by relatively high $\bar{r}$ values (phosphorus-added aluminum-killed steels), and the precipitation hardening type (titanium-added steels) of high yield strength, which are both produced by continuous- and batch-annealing processes. Since some information regarding these two types has already been presented elsewhere,[2,6] major attention will be paid to dual-phase steels.

TABLE I

Cold Rolled HSLA of Nippon Steel Corporation;
Typical Mechanical Properties of 0.8 mm Thick Sheets

Grade in Tensile Strength		Continuous Annealing (CAPL)															Batch Annealing				
		Ductility Type (high $\bar{r}$ or low YS high EL)										Precipitation Hardening					Solid Solution Hardening or Precipitation Hardening				
		Solid Solution Hardening					Dual Phase Hardening														
MPa	(ksi)	YS MPa	TS MPa	EL %	YS/TS	$\bar{r}$	YS MPa	TS MPa	EL %	YS/TS	$\bar{r}$	YS MPa	TS MPa	EL %	YS/TS	$\bar{r}$	YS MPa	TS MPa	EL %	YS/TS	$\bar{r}$
400	(58)	270	413	36	0.66	1.38											266	415	37	0.64	1.68
450	(65)	302	450	33	0.67	1.40											299	455	34	0.66	1.50
500	(73)						330	510	34	0.64	1.12	415	514	26	0.80	1.07	399	521	27	0.76	1.14
550	(80)						368	554	33	0.66	1.06	452	556	23	0.81	1.08	419	568	24	0.74	1.11
600	(87)						353	657	28	0.53	0.88	509	625	20	0.81	1.00	483	636	20	0.76	1.10
800	(116)						468	864	22	0.54	0.95										
1000	(145)						775	1040	12	0.75	0.89										

Areas enclosed by broken lines indicate features of each category

There are two typical continuous-annealing cycles in obtaining dual-phase steels.[7] One employs rather mild cooling rates (5-20 degrees C/second), and the other involves drastic quenching, after annealing in the alpha-gamma temperature range. Figure 1 indicates the relationship between tensile strength and total elongation with regard to the dual-phase groups, consisting of mildly cooled ("dual-phase hardening (CAPL)"), and drastically cooled ("water quenched" and "water quenched and tempered") ones, together with other categories of HSLA steels. It is seen that elongation of mildly cooled dual-phase group is superior to any others, including the drastically cooled dual-phase groups, with reference to the strength spectrum.

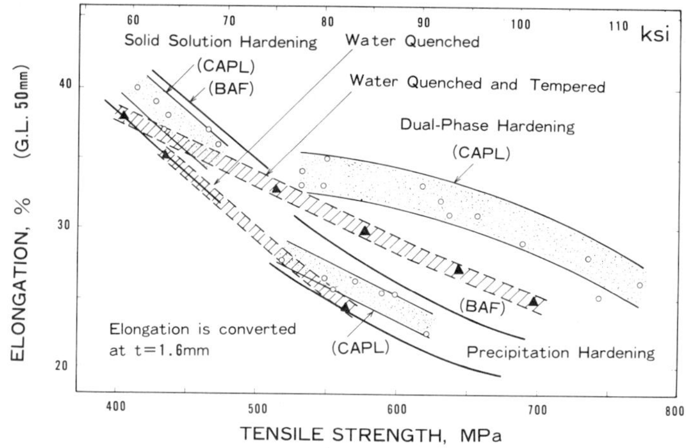

Fig. 1-Relationship between tensile strength and elongation for various cold-rolled HSLA steels.
CAPL: Treated with continuous annealing and processing line.[2,6]
BAF : Treated in batch annealing furnace.

In order to study the cause of the difference in elongation between the mildly cooled and drastically cooled dual-phase steels, some experiments have been carried out. Figure 2 shows the materials and experimental procedure. Steels I and II are dual-phase steels having higher manganese contents and killed with aluminum, Steel III is a phosphorus-added aluminum-killed steel, being an example of the solid solution hardening, high $\bar{r}$ type. Steel IV is a press-formable, aluminum-killed steel. Steels III and IV are included for the purpose of reference. All the steels were heat treated as described in Figure 2 after cold rolling with a 70% reduction, and tensile tested. Attention was paid to mechanical property changes due to the variation of cooling rate after soaking at the alpha-gamma temperatures.

CHEMICAL COMPOSITION (Wt. %)

Steel	C	Si	Mn	Others	Thickness (mm)
I	0.09	0.44	1.54	—	0.8
II	0.08	0.26	2.04	—	0.8
III	0.07	0.02	0.32	P:0.124	0.8
IV	0.04	0.02	0.25	—	0.8

Cold-Rolled Aluminum killed Steels

PROCEDURE

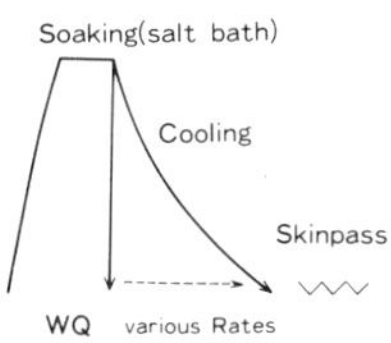

Soaking : 690 to 870°C, 1to2min.

Cooling rate : 0.5 degC/sec to WQ

Skinpass : 0 or 1 %

Figures 3-6 are the experimental results showing the cooling rate dependence of the mechanical properties. It should be noted that the change of elongation with variations in cooling rate is similar for all the steels, with the elongation exhibiting a marked decrease at around 30 degrees C/second. Note also that steel II, having 2% manganese (Figure 4), exhibits low YS/TS values (< 0.5) over a very wide range of cooling rate.

Fig. 2(left)-Materials and experimental procedure for cold-rolled HSLA steels.

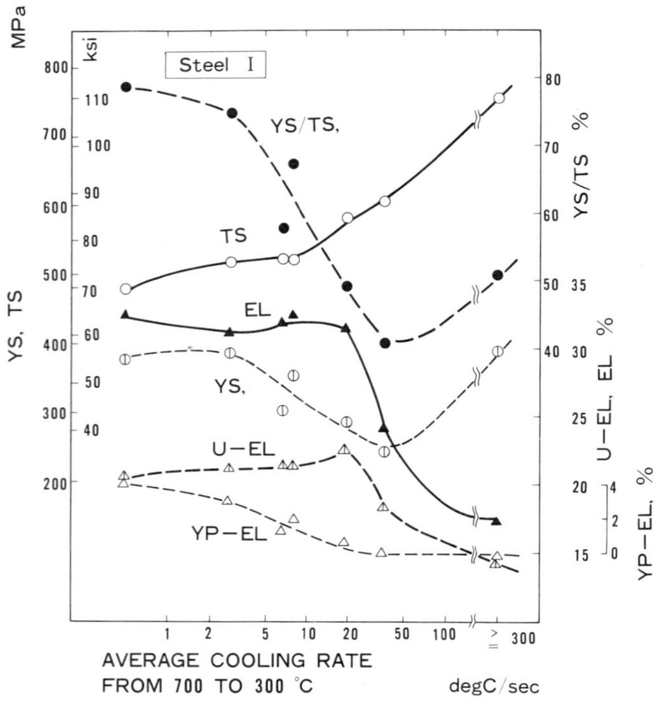

Fig. 3-Cooling rate dependence of mechanical properties of steel I (0.4% Si, 1.5% Mn), after heating at 750°C for 2 min. Not skinpassed.

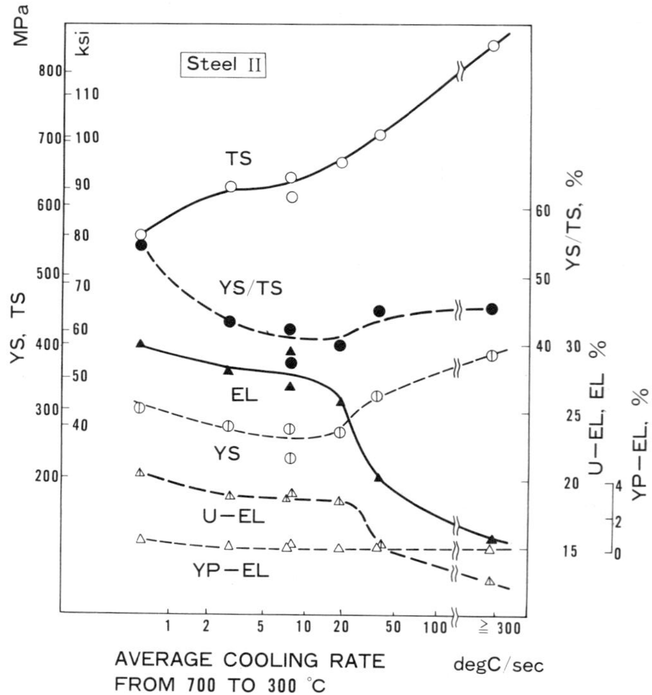

Fig. 4-Cooling rate dependence of mechanical properties of steel II (0.3% Si, 2% Mn), after heating at 750°C for 2 min. Not skinpassed.

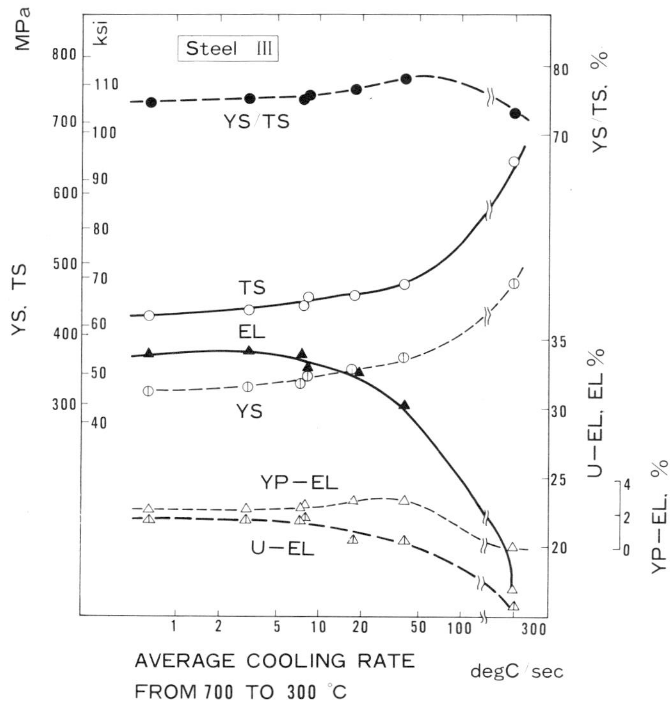

Fig. 5-Cooling rate dependence of mechanical properties of steel III (0.3% Mn, 0.1% P), after heating at 800°C for 2 min. 1% skinpassed.

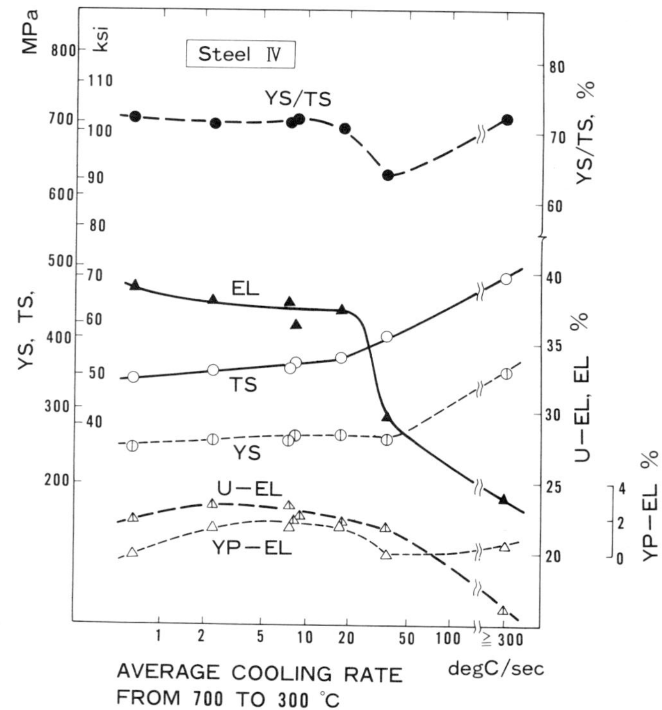

Fig. 6-Cooling rate dependence of mechanical properties of steel IV (0.25% Mn), after heating at 800°C for 2 min. 1% skinpassed.

Figure 7 shows the work-hardening rate, $\varepsilon/\sigma \cdot d\sigma/d\varepsilon$, at each level of strain,* for each steel subjected to cooling rates A (<10^0 degrees C/second), B (~10^1 degrees C/second) and C (water quench). In all cases, water quenching (C) gave the lowest level of the work-hardening rate, and poorest uniform elongation. The dual-phase steels, I and II, show a very high work-hardening rate at small strains and do not follow the power law (i.e., n is not considered to be constant in a wide range of ε). On the other hand, the reference steels III and IV exhibit an almost constant work-hardening rate with a nearly constant n value.

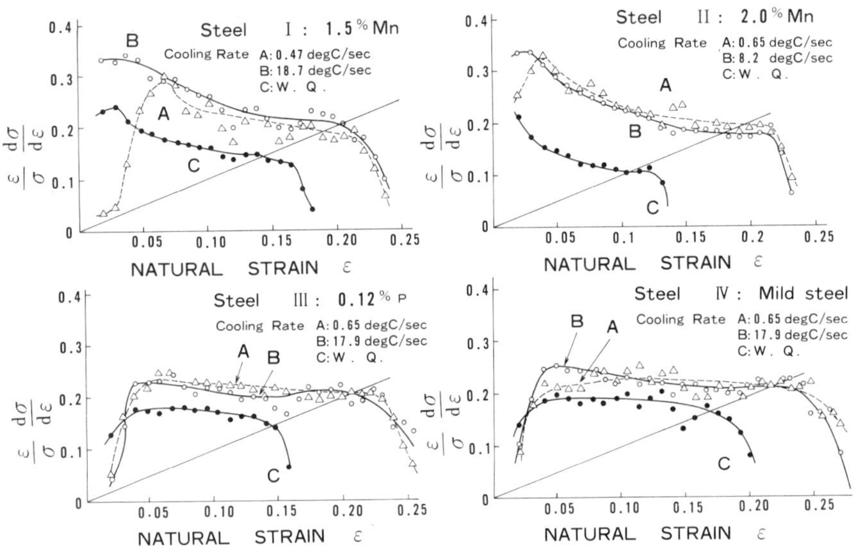

Fig. 7-Dependence of work-hardening rate on strain of cold-rolled steels, under various conditions of cooling rate after annealing.

In steel II, all the cooling rates A, B and C give rise to low YS/TS values (Figure 4). However, there are some differences in microstructure depending on the cooling rate, as shown in Figure 8. A detailed study concerning microstructure is now underway. Preliminary results using both optical and scanning electron microscopy indicate the following changes in microstructure with cooling rate. At the slowest cooling rate, A, the microstructure consists of ferrite, fine pearlite and a small amount of martensite. At the intermediate cooling rate, B, the microstructure changes to ferrite and martensite islands of about 15% by volume, and is replaced by ferrite and massive martensite of about 35% by volume at the fastest cooling rate, C.

As seen in Figure 7, the work-hardening rate is higher for the mildly cooled cases than for the drastically quenched ones, regardless of whether the chemistry is for dual-phase steel (steels I and II) or not (steels III and IV). This suggests that the mildly cooled steels have a cleaner ferrite matrix with respect to solute carbon (nitrogen will be of minor importance since all the steels are aluminum-killed), compared to the drastically quenched ones.[8,9] Figure 9 illustrates the results of an internal friction study on steels I and IV which supports the view that there is less carbon in solution in the

* See appendix.

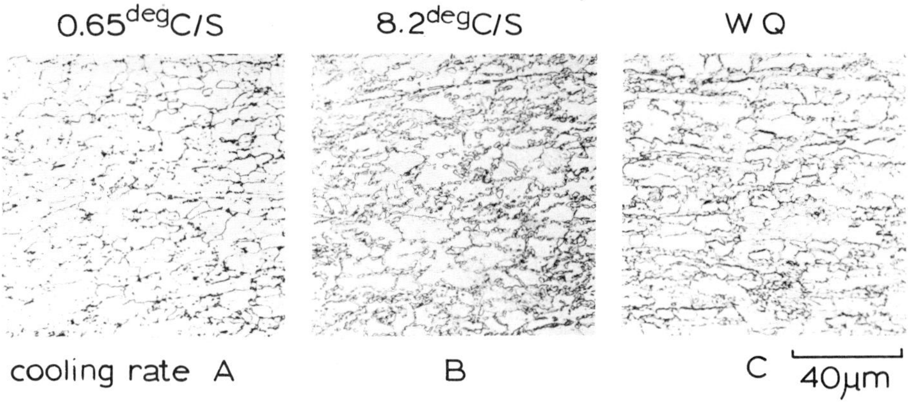

Fig. 8-Microstructures of steel II (0.3% Si, 2% Mn), obtained by cooling at various rates after annealing.

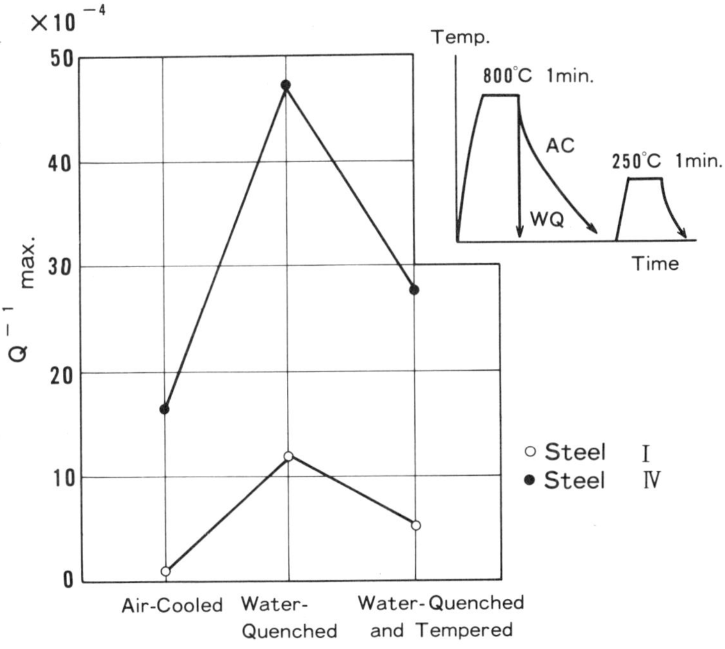

Fig. 9-Effect of cooling rate on internal friction.

air-cooled steels (about 8 degrees C/second) compared with the water quenched, or water-quenched and tempered steels.

Figure 10 shows the variation in tensile properties with soaking (annealing) temperature for a dual-phase steel (II) after mildly cooling and water quenching after soaking. It can be seen that both the tensile and yield strengths of the mildly cooled steel exhibit little variation in the temperature range of 750 to 870°C, combined with a good level of elongation. On the contrary, water quenching can be seen to produce wide variations in tensile properties with soaking temperature.

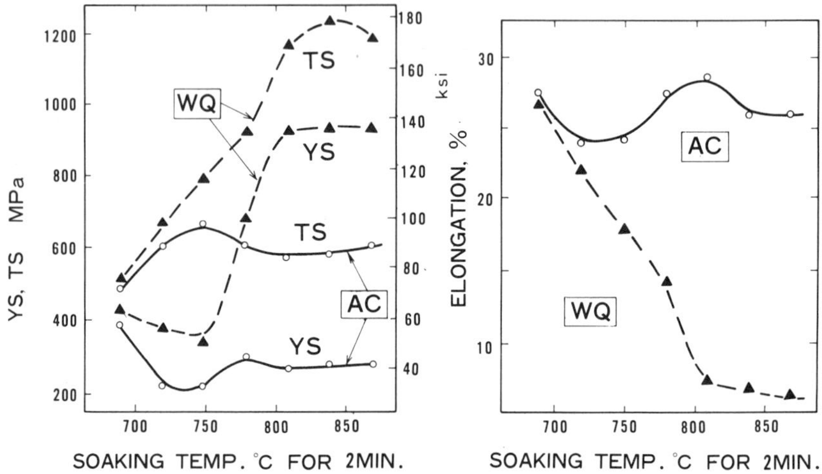

Fig. 10-Effect of soaking temperature on mechanical properties of steel II (0.3% Si, 2% Mn) after air cooling or water quenching.

3. Ultra-High-Strength Grade-A Variation of Cold-Rolled Dual-Phase Steel

This is a peculiar version of dual-phase steel, consisting of unrecrystallized ferrite and martensite. An example of material, having 2% manganese and a small amount of titanium, is presented in Figure 11 where the production process is also described. The important point is the controlled annealing technique to produce uniformly recovered ferrite structure, which is attained only by continuous annealing.

Figure 12 shows the mechanical properties as a function of the continuously annealing temperature. Recovery of ferrite takes place from about 650°C. Recrystallization commences at about 750°C. This high temperature of recrystallization may be due to the existence of fine precipitates of titanium carbides. In the temperature range of 700 to 740°C, austenite formation occurs, although the ferrite matrix is still in the recovery stage. The austenite very likely forms from the pearlite phase present before annealing. After cooling from this temperature range (e.g., 730°C), the peculiar structure is brought out, showing a decrease in the yield strength with a high level of tensile strength, indicating the transformation of the austenite to martensite. Figure 13 shows two optical microstructures of this steel, annealed for one minute at 670°C and 730°C respectively.

The mechanical properties obtained are shown in Table II, and exhibit good combinations of strength and ductility with 1000 MPa in tensile strength and 10% or more total elongation.

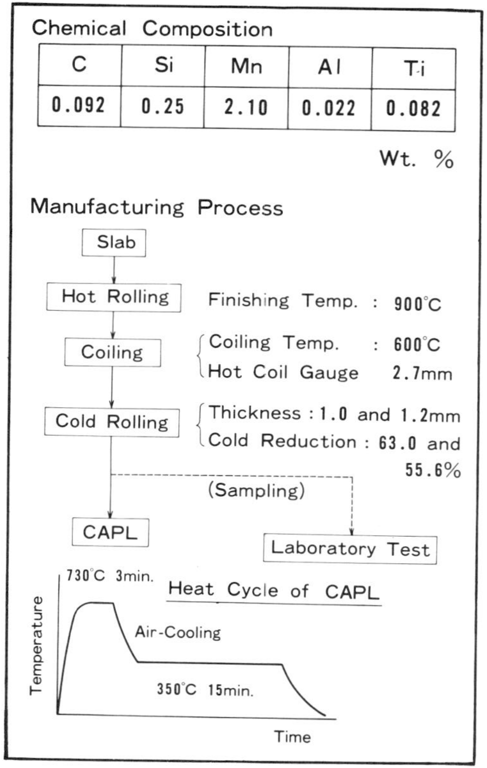

Fig. 11-Material and production process for ultra high-strength grade.

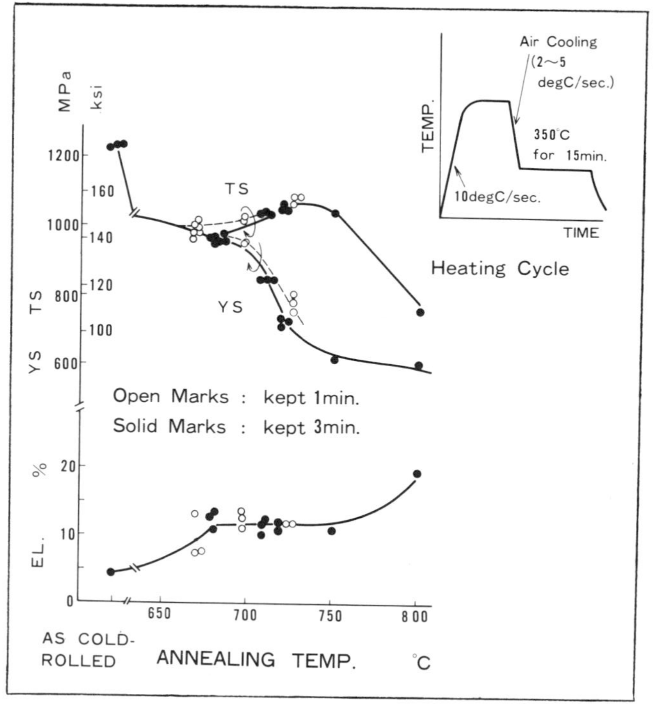

Fig. 12-Effect of continuously annealing temperature on mechanical properties of 2% Mn, 0.08% Ti cold-rolled steel.

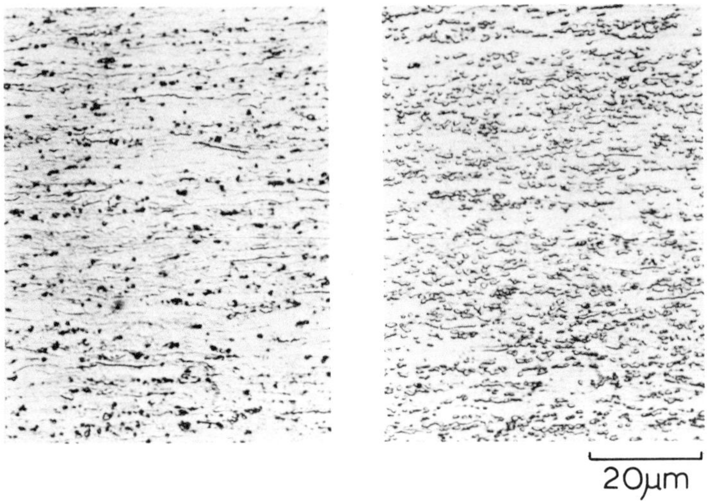

Fig. 13-Microstructures of 2% Mn, 0.08% Ti cold-rolled steel; (a) recovered ferrite and pearlite, (b) recovered ferrite and martensite.

TABLE II

Mechanical Properties of 1000 MPa-TS Sheets Produced by Continuous Annealing (CAPL)

Thickness mm	Direction	YS MPa (ksi)	TS MPa (ksi)	EL %	YP-EL* %	YS/TS
1.2	Longitudinal	755 (110)	988 (143)	12	0	0.76
	Transverse	851 (123)	1011 (147)	11	0	0.84
1.0	Longitudinal	770 (112)	1035 (150)	12	0	0.74
	Transverse	928 (135)	1108 (161)	9	0	0.84

G.L. 50 mm * Yield-point elongation

4. Hot-Rolled Dual-Phase Steels

We have succeeded in producing 500 MPa tensile grade dual-phase hot-rolled gages as well as higher strength grades in commercial mill-trials. Examples of the materials, dual phase treatments and mechanical properties are summarized in Table III. All the steels show YS/TS ratios of about 0.5 or less. The lower tensile grades having smaller amounts of silicon (e.g., steel A) are produced by employing somewhat larger cooling rates after soaking, although the cooling rate is still in the "mild cooling" category. All these steels show better combinations of tensile strength and total elongation compared to the conventional HSLA steels as shown in Figure 14.

Figure 15 shows how the heat treatment affects the tensile properties. The dual-phase heat treatment lowers the yield strength and raises the tensile

TABLE III

Heat Treatment and Mechanical Properties of Hot-Rolled Dual-Phase Steels

Steel	Composition %			Heat Treatment		YS		TS		YS/TS	EL
	C	Si	Mn	Soaking	Cooling Rate	MPa	(ksi)	MPa	(ksi)		%
A	0.08	0.4	1.5	800 °C 2 min	12 deg C/S	270	(39.1)	530	(77.0)	0.51	31.6
B	0.06	1.0	1.5	800 °C 3 min	5 deg C/S	259	(37.6)	533	(77.4)	0.48	35.1
C	0.10	1.1	1.4			338	(49.1)	647	(93.9)	0.52	30.2
D	0.10	1.1	1.6			328	(47.7)	676	(98.2)	0.48	28.0
E	0.11	1.3	1.6			337	(48.9)	769	(111.7)	0.44	25.0
F	0.12	1.4	1.6			372	(54.0)	840	(121.9)	0.44	22.6

2.5 mm thick G.L. 50 mm

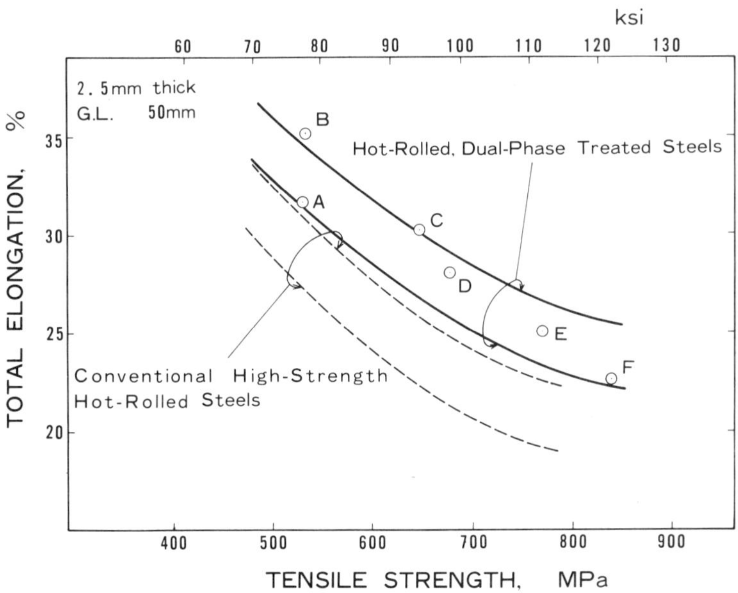

Fig. 14-Tensile strength-elongation balance of dual-phase hot-rolled steels compared with conventional high-strength hot rolled steels.

strength without lowering the good ductility (and even increases the ductility in the higher silicon steel) obtained in the as-hot-rolled condition. These results are in contrast to those of Rashid[3,4] who found that after dual-phase treating a vanadium-containing steel, the tensile strength was the same as the as-rolled condition but the elongation was increased. It is believed that this difference in behavior stems from the influence of precipitates.

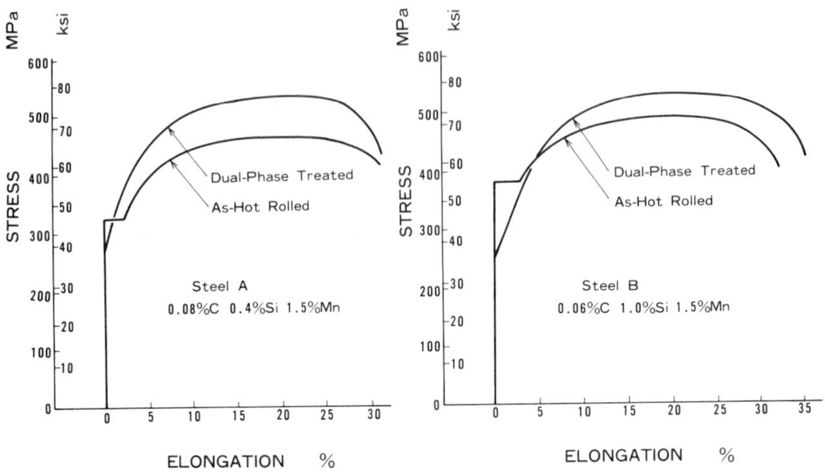

Fig. 15-Effect of dual-phase heat treatment on tensile properties of hot-rolled steels.

In the case of the vanadium-bearing steel, the presence of vanadium carbonitrides (VC,N) in the as-rolled state will lead to high tensile and yield strengths with rather poor elongation. When heating into the alpha-gamma temperature range, the VC,N dissolves. After cooling, the VC,N probably does not precipitate too much, remaining mostly in solution in the ferrite; thus the ferrite matrix is "cleaned" in the sense of being "precipitate-free" and consequently softer and more ductile. If coincidentally, the increase in tensile strength produced by the newly developed martensitic phase is equal to the decrease in tensile strength of the now "clean" ferrite, then the net tensile strength after the dual-phase treatment will be the same as the as-rolled state. The elongation, however, will be increased because of the more ductile state of the ferrite.

The fact that the present steels show little change in elongation but increases in tensile strength after the dual-phase treatment can be explained on the basis that as the ferrite is essentially "precipitate-free" in the as-rolled state, the changes in the ferrite properties after heat-treatment are relatively smaller in magnitude than in the vanadium-bearing steel. Consequently, the tensile strength increase produced by the martensite formation is not counter-balanced by any ferrite softening effect; neither is the overall elongation changed from the originally good as-hot-rolled values. It is possible that the present steels have an advantage over vanadium-bearing steels in that they may be able to attain "dual-phase" properties with shorter heat-treatment times. This is expected because the rate controlling factor is the dissolution of cementite, rather than the dissolution of VC,N, which is likely to be a much more rapid process. It would be worthwhile to study the effect of precipitation on dual-phase treated steels more closely.

Figure 16 shows the work-hardening rate versus strain for as-hot-rolled and dual-phase treated steels. It is noticed that steel B containing higher silicon attains a very high work-hardening rate in the dual-phase state especially at low strain levels. Silicon is considered to drive carbon atoms from the ferrite matrix into the austenite phase[10] producing a more stable austenite as well as a cleaner ferrite when subjected to the heat treatment,

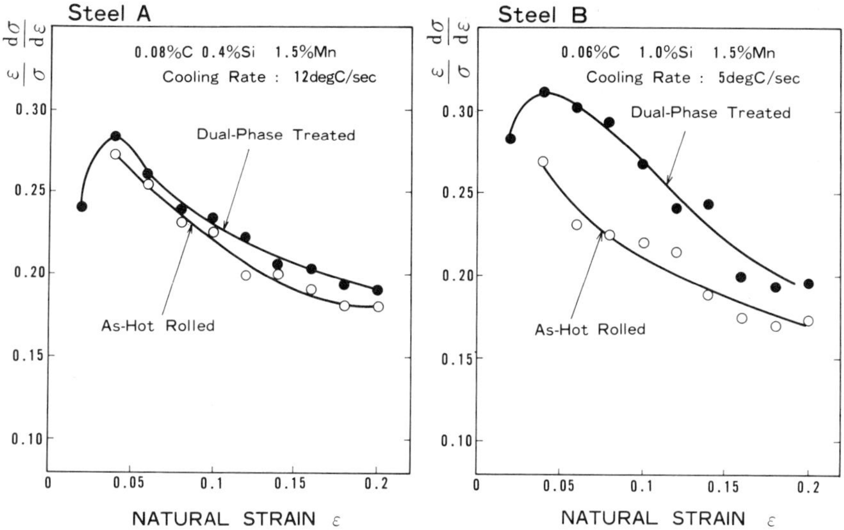

Fig. 16-Dependence of work-hardening rate on strain of hot-rolled steels.

and also suppresses the formation of pearlite on cooling.[11] Presumably these are the reasons why steel B is improved both in strength and in elongation. Steel B has another advantage for easier production because slower cooling is available to attain the dual-phase structure.

5. Summary

In addition to solution- and precipitation-hardening HSLA steels, both hot- and cold-rolled dual-phase sheet steels can now be produced on a commercial basis. The compositions of the dual-phase steels are based upon plain carbon silicon-manganese steels, without using any particular carbo-nitride forming elements.

It has been found that after the dual-phase heat treatment, a cooling rate less than about 30 degrees C/second is preferable to drastic quenching in order to ensure good elongation by minimizing the amount of carbon in solution in ferrite matrix. By employing this cooling rate, silicon-manganese steels retain or even improve their inherent good elongation, with the simultaneous attainment of a decrease in yield strength and an increase in tensile strength, after the dual-phase treatment.

A controlled continuous annealing technique applied to a cold-rolled silicon-manganese-titanium steel produces an extra high-tensile dual-phase steel with a recovered ferrite matrix and martensite particles, having a good combination of strength and ductility.

Appendix

(1) In the well-known power law equation $\sigma = K\varepsilon^n$, it is assumed that n=const. for small variation range of ε.

Then $d\sigma/d\varepsilon = n\sigma/\varepsilon$, so that

$$(\varepsilon/\sigma \cdot d\sigma/d\varepsilon)_{\varepsilon_i} = n_i$$ i.e. work-hardening rate at ε_i.

(2) Condition giving uniform elongation is

$$dp=0$$

Since $P=\sigma A$, the condition is

$$Ad\sigma+\sigma dA=0,$$ thereby

$$-dA/A=d\sigma/\sigma. \quad \ldots 1$$

Since $A_0=(1+\varepsilon)A$, for $\varepsilon \ll 1$

$$A_0 \simeq e^{\varepsilon}A.$$

Differentiating,

$$Ad\varepsilon+dA=0 \quad \ldots 2$$

From 1 and 2.

$$d\sigma/d\varepsilon=\sigma, \text{ or } \varepsilon/\sigma \cdot d\sigma/d\varepsilon=\varepsilon.$$

P : applied load, σ: stress, A: cross section area,
A_0: original cross section area, ε: strain.

References

1. J.J. Jurkowski and R.N. Keller: SAE Paper 770215, 1977.

2. S. Hayami and T. Furukawa: Proceedings of Microalloying 75, 2A, Washington D.C., 1975, pp. 78-87.

3. M.S. Rashid: SAE Paper 760206, 1976.

4. M.S. Rashid: SAE Paper 770211, 1977.

5. J.H. Bucher and E.G. Hamburg: SAE Paper 770164, 1977

6. K. Toda, H. Gondoh, H. Takechi and H. Masui: Met. Trans., 1976, Vol. 7A, pp. 1629-1636.

7. K. Araki, K. Nakaoka, M. Abe and N. Ohashi: 9th Biennial Conf. of IDDRG Proceedings, 1976, pp. 39-53.

8. W.C. Leslie, G.B. Williams and D.S. Hutton: JISI, 1961, Vol. 198 pp. 21-29.

9. H. Masui, M. Kawaharada and H. Takechi: TETSU-TO-HAGANE, 1972, Vol. 58, pp. 1096-1106.

10. P. Abramowitz and R.A. Moll: Met. Trans., 1970, Vol. 1, pp. 1773-1775.

11. R.F. Hehemann: Phase Transformations, p. 410, ASM, Metals Park, Ohio. 1970.

STRUCTURE AND PROPERTIES OF COLD-ROLLED FERRITE-MARTENSITE (DUAL-PHASE) STEEL SHEETS

P. R. Mould and C. C. Skena

Research Laboratory
United States Steel Corporation
Monroeville, Pennsylvania

Abstract

This study confirms previous published work that dual-phase (ferrite-martensite) steels can be developed in slowly cooled (about 25°F/hr) box-annealed cold-rolled sheets of steels containing at least 2.5 percent manganese provided intercritical annealing temperatures of at least 1300°F are used. Tensile properties were developed that are characteristic of dual-phase steels—that is, low yield strengths (about 30 to 35 ksi), high tensile strengths (about 73 to 86 ksi), good elongations (25 to 29%), low ratios of yield/tensile strength, and an absence of yield-point elongations. For higher manganese steels (3.35%), the yield and tensile strengths were increased to about 40 and 108 ksi, respectively. Lower manganese steels (1.5%) did not develop dual-phase microstructures after slow (box anneal) cooling, but did so when the cooling rate was increased to 3000°F/hr (which approaches air cooling rates). In addition, the faster cooling rate increased the yield and tensile strengths of the dual-phase 2.5 and 3.35 percent manganese steels.

Scanning- and electron-microscope studies of the dual-phase steels confirmed the presence of martensite which was often associated with pearlite and/or bainite areas. A high dislocation density was observed in the ferrite immediately adjacent to martensite areas which, together with the presence of residual stresses in the ferrite, probably accounts for the yielding characteristics of the steels. The volume fraction of martensite (f_m) was found to correlate closely with the tensile strength; f_m values of about 0.2 appear necessary to achieve a tensile strength of 100 ksi.

Low intercritical annealing temperatures (1300 to 1340°F) were found to produce the low ratios of yield/tensile strength that lead to good formability in press-forming operations. However, at high intercritical annealing temperatures (1420°F), higher ratios of yield/tensile strength were obtained which resulted from a decreased proportion of martensite and which are indicative of lower formability. The low volume fraction of martensite observed after annealing at a high temperature is explained on the basis of transformation of a low-carbon austenite formed at the high annealing temperature.

Introduction

The recent automotive demand for lightweight materials to improve gasoline mileage has led to the development of new high-strength steel sheets. Because the formability of most steels decreases as the strength increases, much of the development of high-strength steels has concentrated on optimizing formability. One approach to improving formability in high-strength steels is the production of steels that have microstructures consisting essentially of ferrite together with some martensite (pearlite and/or bainite may also be present). These steels are often referred to as dual-phase steels. Compared with the more usual ferrite-pearlite steels, the presence of martensite in dual-phase steels modifies the tensile characteristics. For example, continuous yielding (absence of a yield-point elongation) occurs and higher work-hardening rates, tensile strengths, and total elongations are obtained. The advantages of ferrite-martensite steels and means for producing them by intercritical heat treatments have been discussed by Grange.[1*)] In addition, ferrite-martensite steels are used to advantage in modern line-pipe steels where increased strengths are obtained during final expansion of the formed pipe as a result of the continuous-yielding behavior and high work-hardening rate of the steels.[2,3)]

The use of dual-phase steels for many automotive sheet applications has been promoted by GM Research workers who have contributed importantly to their development.[4,5)] For cold-rolled sheets, Matsuoka and Yamamori[6)] showed that ferrite-martensite steels can be obtained even after the very slow cooling associated with box annealing, provided that high-manganese (about 2.5%) steels and intercritical annealing temperatures are used. Since the work of Matsuoka and Yamamori, leaner steel compositions have been developed for cold-rolled sheets.[7)] Development of the dual-phase structure in such steels requires that the steel be cooled relatively rapidly (as in continuous annealing, for example) from the intercritical annealing temperature.

Because of limited continuous-annealing facilities in the domestic steel industry for developing dual-phase microstructures in either hot- or cold-rolled sheets, there is interest in using other existing facilities (such as box or open-coil annealing facilities) to produce the dual-phase steels. Thus the purpose of the present work was to further explore earlier observations that high-manganese steels can develop dual-phase microstructures during the very slow cooling (about 25°F per hour) that is associated with box annealing. More specifically, efforts were made to determine whether dual-phase structures could be developed in box-annealed steels containing as low as 1.5 percent manganese. Such steel is more suited to current steelmaking practices. Furthermore, to establish whether additional strength could be derived by using manganese contents above 2.5 percent, steels containing 3.35 percent manganese were also included in this study. An important objective of the current work was the study of the effect of the intercritical annealing temperature and the subsequent rate of cooling on the microstructure and mechanical properties of the manganese steels after they had been cold-reduced on a cold-rolling mill.

* See References.

Materials and Experimental Work

Steel Compositions

For the present study, three 100-pound experimental heats were melted and poured in vacuum into 3- by 8- by 14-inch ingot molds. The compositions of the steels are shown in Table I. All the steels had a nominal carbon content of 0.05 percent, and the manganese contents increased from 1.5 percent for steel A to 2.5 and 3.35 percent for steels B and C, respectively. Steel A contained 0.26 percent silicon, whereas the other steels had residual silicon contents less than 0.03 percent.

Table I

Composition of Experimental Steels Studied

Steel Identity	Composition, weight percent								
	C	Mn	P	S	Si	Sol Al	Total Al	N	O, ppm
A	0.043	1.54	<0.001	0.012	0.26	0.041	0.041	0.003	31
B	0.057	2.50	0.001	0.003	<0.03	0.049	0.049	0.006	21
C	0.055	3.35	0.001	0.003	<0.03	0.043	0.047	0.007	22

Although silicon increases the hardenability of steels (and hence increases the capability for forming martensite in dual-phase steels) the additional hardenability arising from 0.26 percent silicon in steel A is small compared with that obtained from additions of manganese above 1.5 percent. Thus the small difference in silicon contents for the three steels was expected to have a very minor influence on the properties of the steels. All three steels were aluminum-killed in the ladle (about 0.045% aluminum), and consequently had low oxygen contents (21 to 31 ppm).

Processing to Cold-Rolled Sheet

The steel ingots were hot-rolled to 1-inch-thick slabs, which were then surface-ground to 0.8-inch thickness, reheated, and hot-rolled to 0.1-inch-thick sheet. Following hot-mill finishing at 1650°F, the sheets were water-spray-cooled to 1000 to 1100°F, and then slowly cooled in a furnace at 75°F per hour to simulate the cooling of commercially produced hot-rolled coils.

The hot-rolled sheets were shot-blasted and pickled to remove surface scale and then cold-rolled to final sheet thicknesses between 0.035 and 0.045 inch.

Heat Treatments

To ensure that the annealing temperatures chosen for the cold-rolled sheets would be in the intercritical range, austenite start and finish temperatures were initially estimated from equations developed by Grange[8)] and subsequently measured by dilatometry. On the basis of these studies, longitudinal tensile blanks sheared from each of the three cold-rolled steels were heated in argon in a tube furnace to annealing temperatures of 1340, 1380, or 1420°F, held for 6 hours, and cooled at two different rates—slow cooling at 25°F per hour and relatively fast cooling at 3000°F per hour. Specimens of steels containing 2.5 and 3.35 percent manganese were also annealed at 1300°F, which is within their measured intercritical temperature range. The 25°F per hour closely simulates the rate of cooling in production box annealing, and 3000°F per hour approaches that of slow air cooling, as on a continuous normalizing or annealing line.

Tension Testing and Metallography

Standard sheet tension-test specimens having a 2-inch gage length were machined from the annealed blanks and tested to fracture on a Baldwin tension tester at a strain rate of 10^{-2} per minute. For specimens exhibiting no yield point, the yield strength was determined as the 0.2 percent offset flow stress.

The microstructures of the steels were examined by optical microscopy (for magnifications up to 1000 times) and by scanning electron microscopy (for magnifications between 1,500 and 12,000 times). Specimens were prepared by mechanically polishing them and etching them in a 20 percent chromic oxide solution which was found to distinguish martensite from pearlite and/or bainite. In some instances, thin foils of selected samples were examined in a Siemens electron microscope at magnifications of 10,000 to 30,000 times.

Results and Discussion

To facilitate discussion, the effects of manganese and of intercritical annealing temperature on the tensile properties and microstructures of the annealed steels are covered separately. In each case these effects are examined first for slow cooling (25°F/hr) and then for fast cooling (3000°F/hr).

Effect of Manganese Content

a) Slow Cooling (25°F/hr)

The effect of manganese content on the tensile properties and microstructure can be assessed by comparing the results obtained for the lowest intercritical annealing temperature used for each of the three steels. Thus the tensile properties of the 1.5, 2.5, and 3.35 percent manganese steels annealed at 1340, 1300, and 1300°F, respectively, are plotted in Figure 1 and shown in Table II. Although only a small change occurred in the yield strength as the manganese content was increased, the ultimate tensile strength increased markedly with the result that the yield/tensile-strength ratio decreased from 0.68 to 0.39 for the range of manganese contents studied. Although the decrease in total elongation that occurred as the manganese increased might be expected to decrease the overall ductility of the steels, the low yield/tensile-strength ratios are expected to distribute strains during biaxial stretching and hence promote relatively good stretch formability.

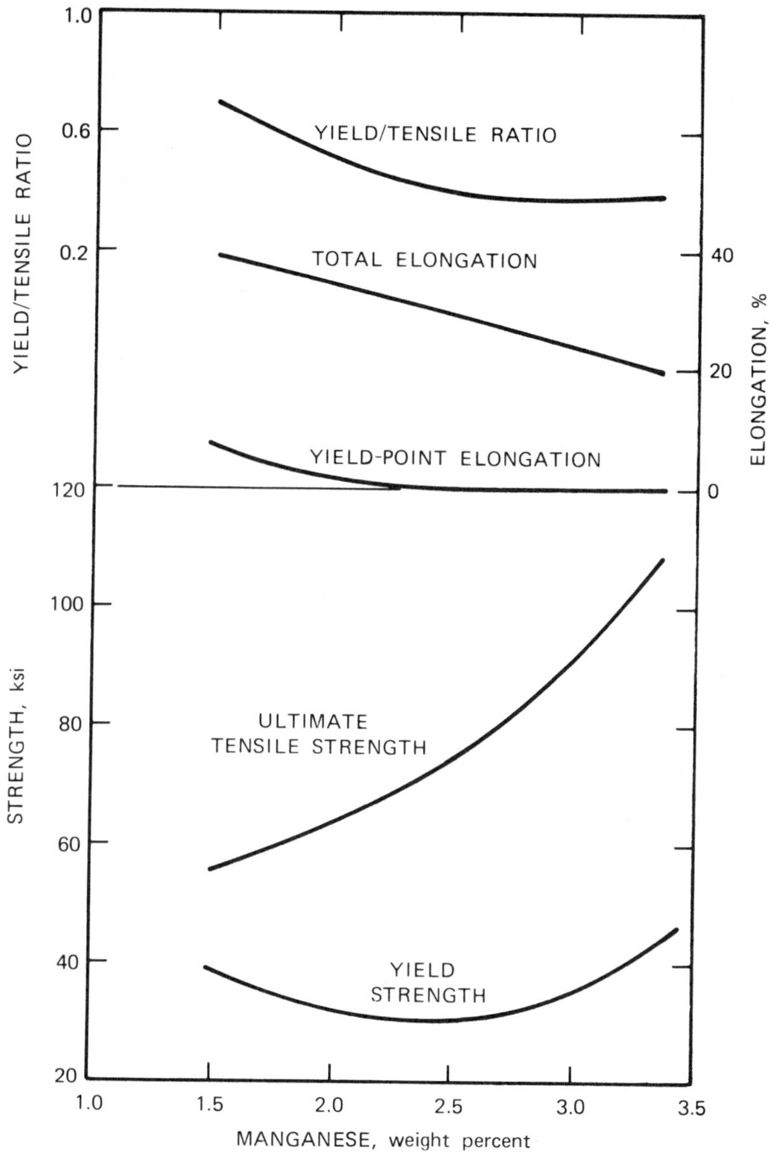

Fig. 1 - Effect of manganese content on tensile properties of steels after slow cooling (25°F/hr) from low intercritical annealing temperatures.

Table II

Longitudinal Tensile Properties of Manganese Steels After Slow Cooling From the Lowest Intercritical Annealing Temperature Studied

Steel	Mn, %	Annealing Temperature, °F	Yield Strength, ksi	Ultimate Tensile Strength, ksi	Total Elongation, %	Yield-Point Elongation, %	Yield Strength / Tensile Strength
A	1.54	1340	38.4	56.3	38.5	6.5	0.68
B	2.50	1300	31.3	74.9	29.0	0	0.42
C	3.35	1300	42.4	108.6	20.0	0	0.39

Furthermore, the high tensile strengths of steels B and C will lead to greater work hardening, compared with steel A, and hence higher strengths can be expected after press forming these steels. Unlike steel A, steels B and C showed an absence of a yield-point elongation, and this behavior together with the high tensile strengths and low ratios of yield/tensile strength is typical of dual-phase steels.[2-7]

The microstructures of steels A, B and C after slow cooling from 1300°F and above are shown in Figure 2. That the microstructure of steel A (1.5% Mn) consisted of ferrite and pearlite demonstrates that the hardenability of this steel was insufficient to cause the formation of a dual-phase structure for cooling rates typical of box annealing (25°F/hr). In contrast, steels B and C exhibited the presence of dark-etching martensite together with patches of relatively fine pearlite. The presence of the martensite in steels B and C accounts for the absence of yield-point elongations and the lower yield strengths and higher tensile strengths in these steels. A comparison of Figures 2b and 2c shows the ferrite grain size of steel C to be somewhat finer than that of steel B. This contributed to the higher yield strength of steel C than for steel B (42.4 vs 31.3 ksi). In addition, the greater volume fractions of pearlite and martensite in steel C (Figure 2c) also probably contributed to its higher yield and tensile strength. Greater amounts of pearlite and martensite (austenite at the annealing temperature) can be expected as a result of a reduction in the eutectoid composition as the manganese is increased.[9] However, the higher tensile strength of steel C probably was most influenced by the higher proportion of martensite in this steel than in steel B.

Electron transmission microscopy of steel B (2.5% Mn) confirmed that the dark-etching patches were martensite. The high contrast (high dislocation density) and areas of internally twinned substructure in the dark patches of the electron micrographs (such as Figure 3) are typical of medium-carbon martensites, which have been shown to contain a mixture of both lath and plate martensite.[10]

Transmission electron microscopy also showed that not all the austenite formed during intercritical annealing subsequently transformed to martensite; some transformed to pearlite. Figure 4 shows that transformation to pearlite and martensite occurred within the same original austenite area. The cusp shape of the pearlite/martensite interface in Figure 4 marks the extent of pearlite growth before the remaining austenite transformed to martensite.

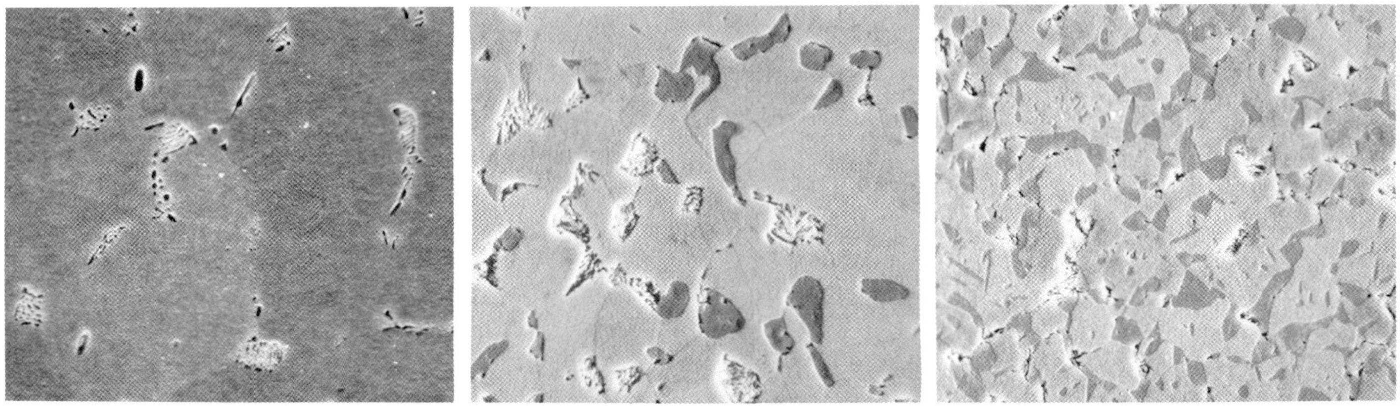

(a)
Steel A (1.5% Mn)
(1340°F)

(b)
Steel B (2.5% Mn)
(1300°F)

(c)
Steel C (3.35% Mn)
(1300°F)

Fig. 2 - SEM micrographs of steels slowly cooled (25°F/hr) from low intercritical annealing temperatures. X3000.

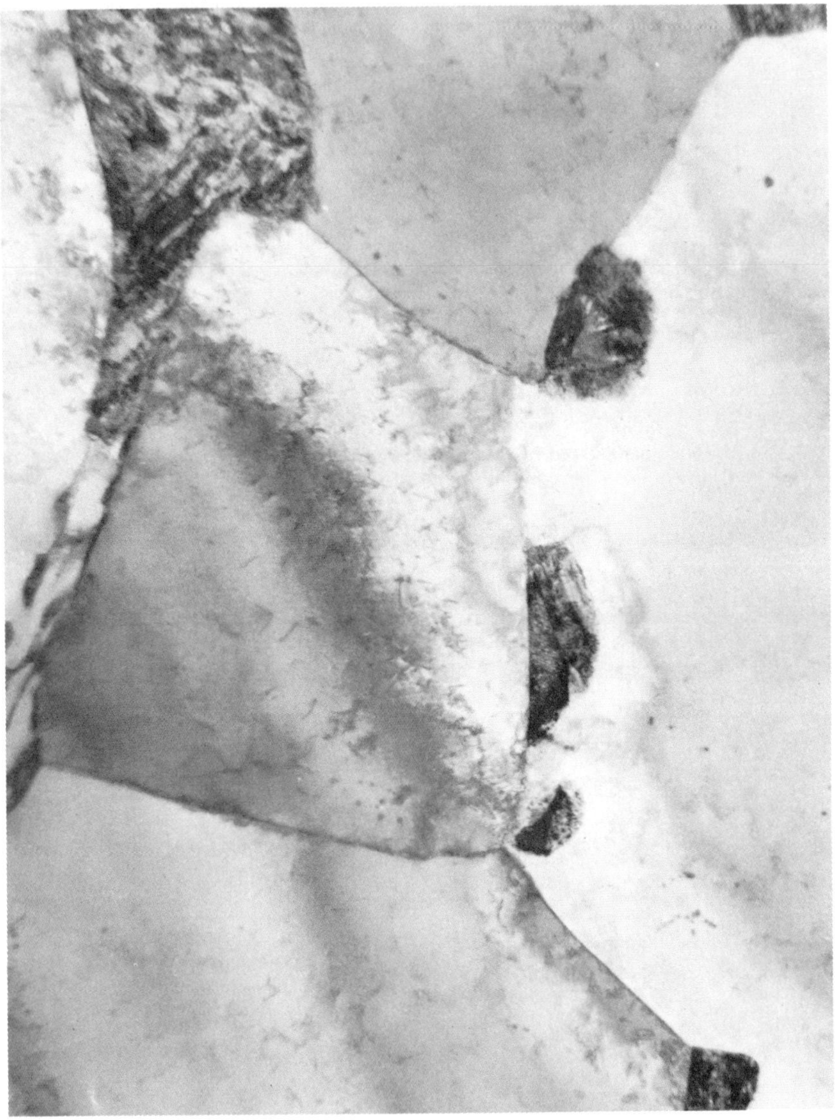

Fig. 3 - Electron micrograph of steel B after slow cooling (25°F/hr) from 1300°F. X15,000.

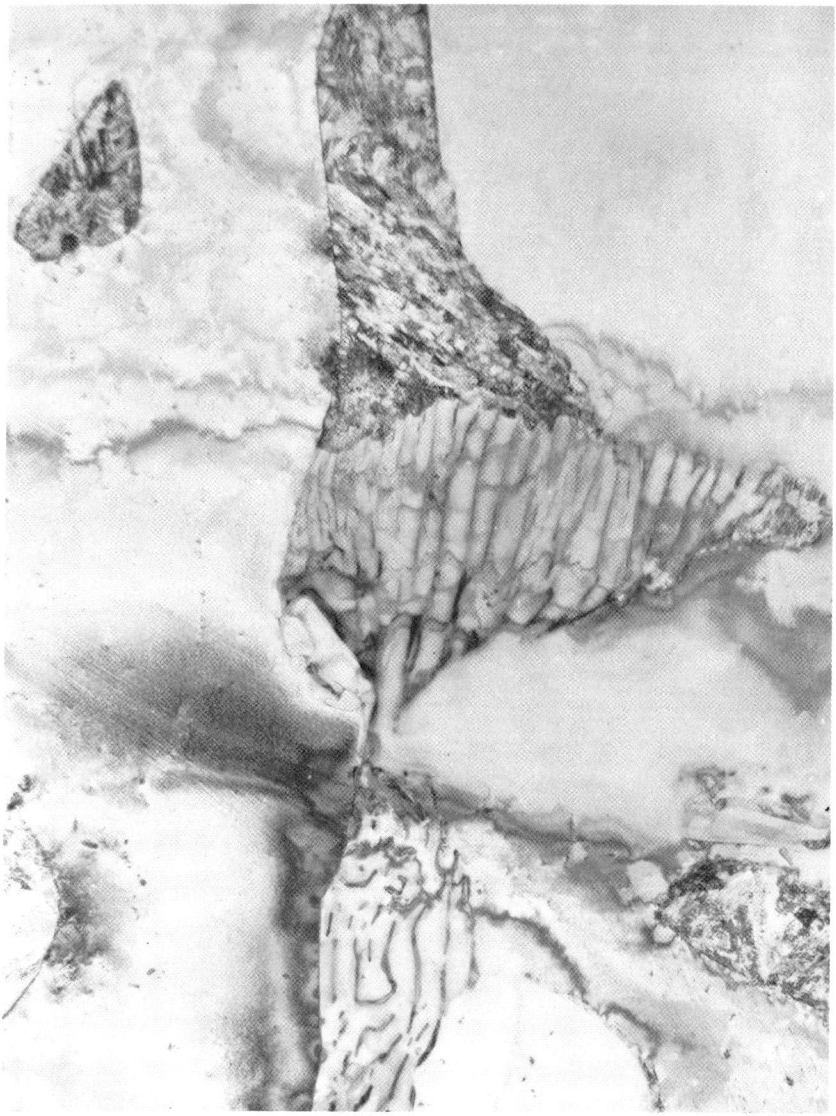

Fig. 4 - Electron micrograph of steel B after slow cooling (25°F/hr) from 1300°F. X15,000.

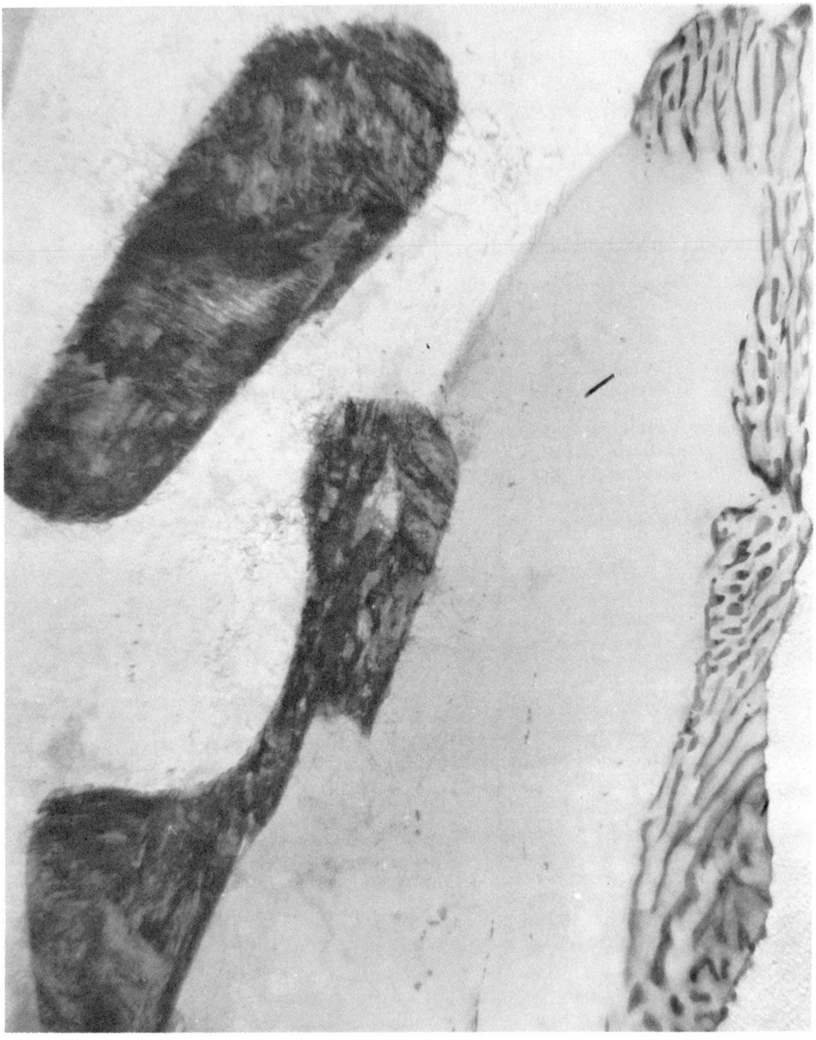

Fig. 5 - Electron micrograph of steel B after slow cooling (25°F/hr) from 1300°F. X15,000.

Figure 5 shows that in some areas of steel B a high dislocation density was present in the ferrite immediately adjacent to the martensite, whereas very few dislocations were observed in the ferrite adjacent to pearlite areas. This observation suggests that the dislocations are generated at the ferrite-martensite interface as a result of the large local strains development by the expansion that accompanies martensite formation. These dislocations are believed to be very mobile and probably account for the continuous-yielding behavior of the steels in a manner analogous to the effect of temper rolling on yielding in low-carbon steel sheets. However, the presence of locked-in residual stresses (both elastic and plastic) probably contributes to the lowering of the yield strength and the rapid work-hardening rate of the martensite-containing steels, as discussed by Dabkowski and Speich.[2)]

b) Fast Cooling (3000°F/hr)

The tensile properties of the experimental steels that were fast-cooled from low intercritical annealing temperatures are shown in Table III and plotted in Figure 6. For comparison, Figure 6 also shows the corresponding properties for samples slow-cooled (box-anneal cooling) from the intercritical temperature (as illustrated in Figure 1). Whereas the yield strength of the slow-cooled steels decreased slightly as the manganese content increased from 1.5 to 2.5 percent (associated with the transition from a ferrite-pearlite structure to a ferrite-martensite-pearlite structure), the yield strength of the fast-cooled steels increased progressively as the manganese increased to 3.35 percent. The tensile strengths of the fast-cooled steels increased, and the elongations decreased, as the manganese content was increased. All the fast-cooled steels showed an absence of a yield-point elongation, and the yield/tensile-strength ratios were low (0.40 to 0.49), indicating that a ferrite-martensite mixture was present. For the most part, the yield strengths and the tensile strengths were higher, and the total elongations lower, for the steels cooled at 3000°F per hour than for the slow-cooled steels (25°F/hr).

Table III

Longitudinal Tensile Properties of Manganese Steels After Fast Cooling From the Lowest Intercritical Annealing Temperature Studied

Steel	Mn %	Annealing Temperature, °F	Yield Strength, ksi	Ultimate Tensile Strength, ksi	Total Elongation, %	Yield-Point Elongation, %	Yield Strength / Tensile Strength
A	1.54	1340	29.5	73.0	29.0	9	0.40
B	2.50	1300	37.4	91.5	22.0	0	0.41
C	3.35	1300	59.0	120.5	18.5	0	0.49

The microstructures of the steels cooled at 3000°F per hour are shown in Figure 7. Unlike the slow-cooled 1.5 percent manganese steel (steel A), which showed a structure of ferrite and pearlite (Figure 2a), the same steel after fast cooling showed a microstructure of ferrite, pearlite, and a dark-etching phase (Figure 7a). The dark-etching phase in Figure 7a was identified as martensite by transmission electron microscopy; for example, the electron micrograph shown in Figure 8 is very similar to those of Figures 3 through 5. Thus, fast cooling of steel A resulted in the partial transformation of austenite to martensite, which in turn accounted for the absence of a yield point, the relatively high tensile strength, and the low yield/tensile-strength ratio (Figure 6). As expected the microstructures of steels B and C (Figures 7b and 7c) also contained martensite. However, by comparing Figures 7b and 7c with Figures 2b and 2c, it is apparent that the fast-cooled steels (3000°F/hr) contained more martensite and more pearlite or bainite than their slow-cooled counterparts, which accounts for the higher yield and tensile strengths of the fast-cooled steels (Figure 6).

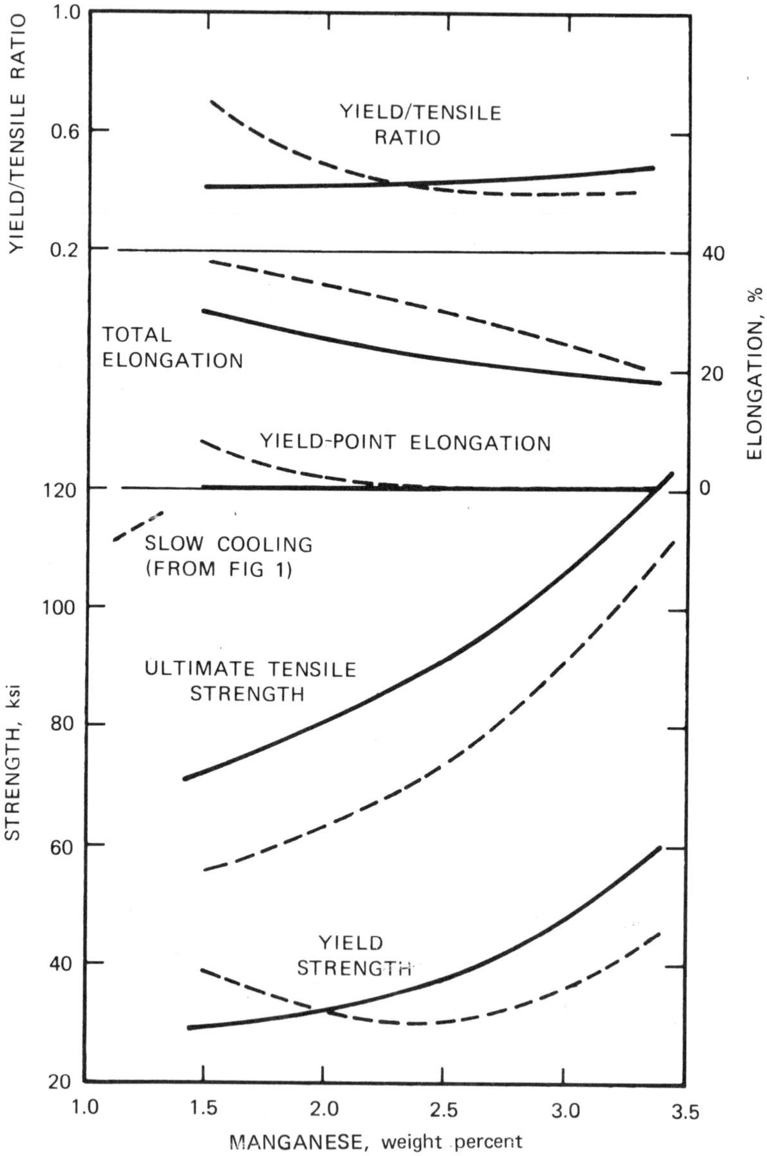

Fig. 6 - Effect of manganese content on tensile properties of steels after fast cooling (3000°F/hr) from low intercritical annealing temperatures.

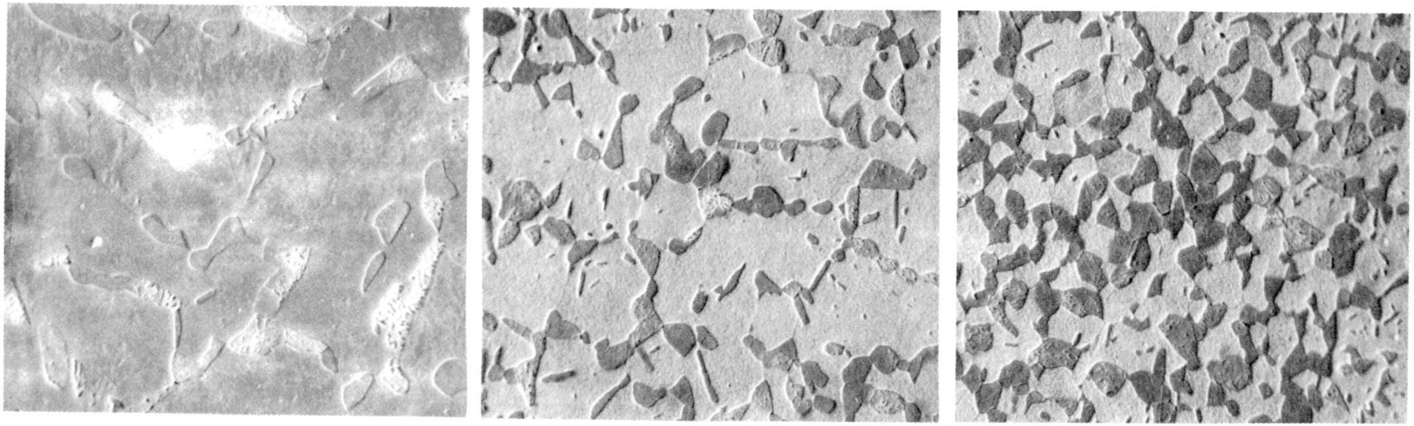

(a)
Steel A (1.5% Mn)
(1340°F)

(b)
Steel B (2.5% Mn)
(1300°F)

(c)
Steel C (3.35% Mn)
(1300°F)

Fig. 7 - SEM micrographs of steels fast-cooled (3000°F.hr) from low intercritical annealing temperatures. X3000.

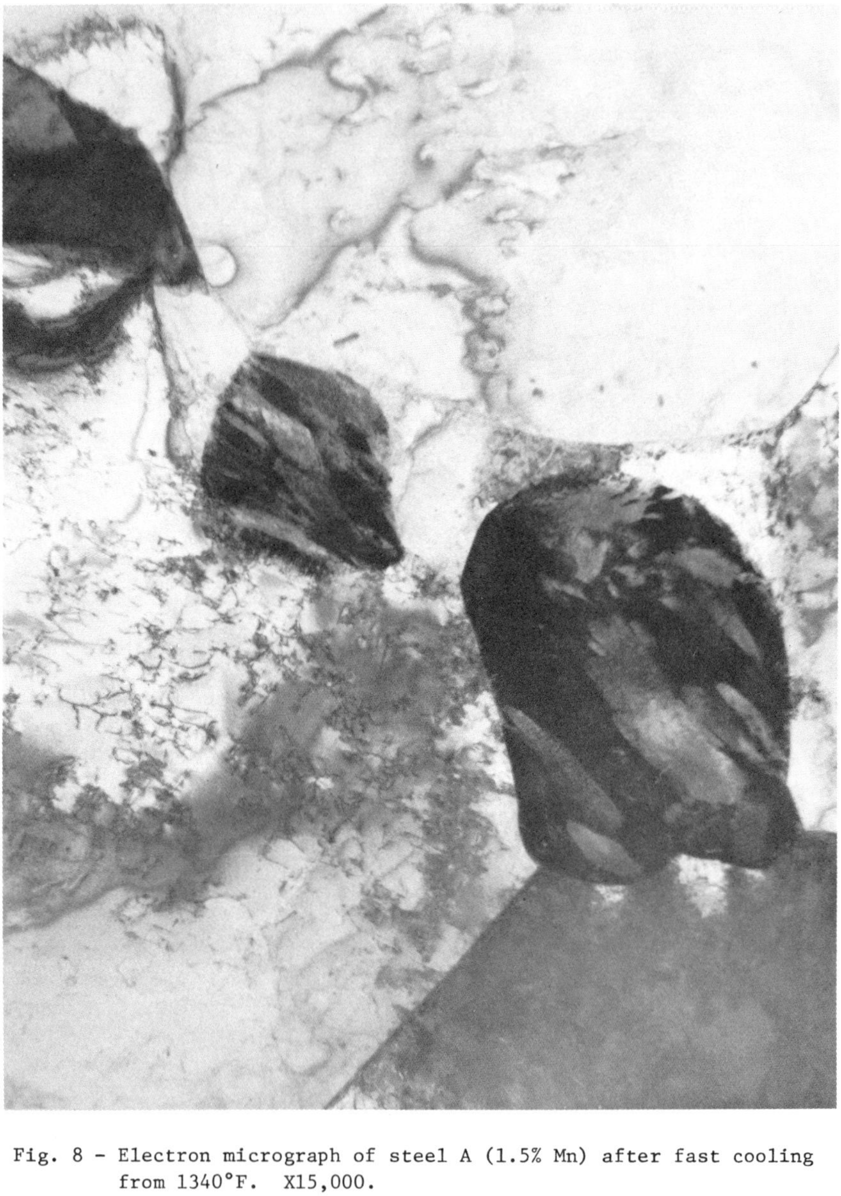

Fig. 8 - Electron micrograph of steel A (1.5% Mn) after fast cooling from 1340°F. X15,000.

Effect of Intercritical Annealing Temperature

a) Slow Cooling (25°F/hr)

Figure 9 shows that the intercritical annealing temperature has a strong influence on the tensile properties of steels B and C, but little influence on those of steel A. This is not surprising in view of the observation that martensite was not developed in steel A after slow cooling from 1340°F. Thus for steel A almost no change in tensile properties occurred as a result of changes in the annealing temperature, and all three annealing temperatures resulted in a significant yield-point elongation for steel A (Table IV), which is a characteristic of annealed steels having ferrite-carbide microstructures. Subsequent microstructural examinations confirmed the presence of ferrite and pearlite and the absence of martensite in all of the intercritically annealed samples of steel A.

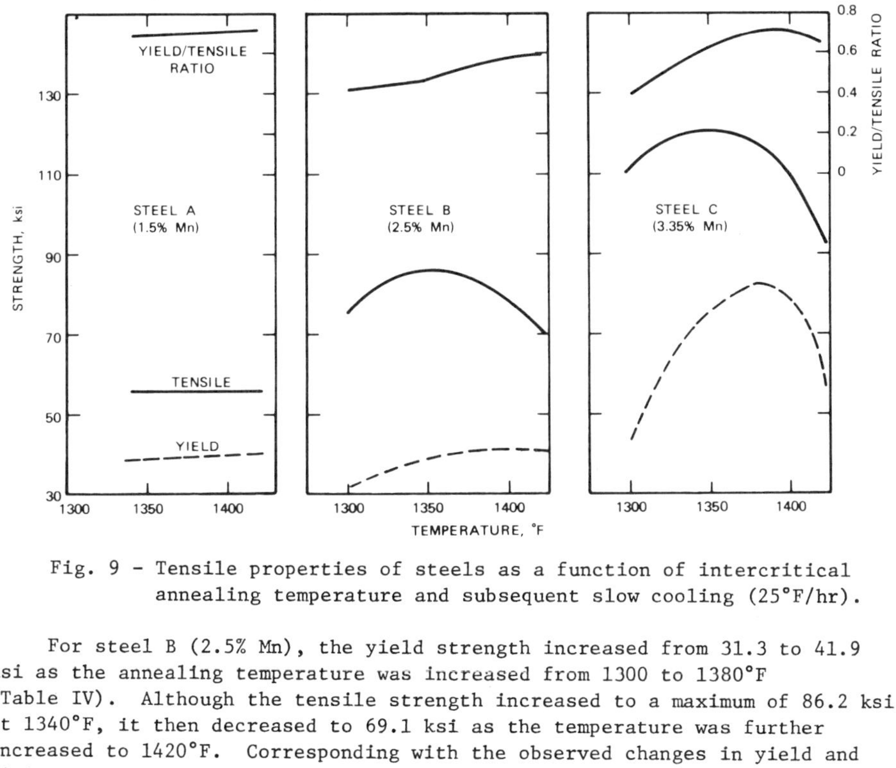

Fig. 9 - Tensile properties of steels as a function of intercritical annealing temperature and subsequent slow cooling (25°F/hr).

For steel B (2.5% Mn), the yield strength increased from 31.3 to 41.9 ksi as the annealing temperature was increased from 1300 to 1380°F (Table IV). Although the tensile strength increased to a maximum of 86.2 ksi at 1340°F, it then decreased to 69.1 ksi as the temperature was further increased to 1420°F. Corresponding with the observed changes in yield and ultimate tensile strengths, a gradual increase in the yield/tensile-strength ratio was observed (Figure 9); consequently, a decrease in the stretch formability of steel B is expected when it is annealed at the higher temperatures.

Figure 9 shows that annealing temperature had a marked effect on the tensile properties of steel C (3.35% Mn). The yield strength of steel C increased to a greater extent than that of steel B as the annealing temperature increased, even though the maximum yield strength occurred at 1380°F for both steels. Like steel B, the maximum tensile strength for steel C occurred

Table IV

Longitudinal Tensile Properties of Steels After Slow Cooling (25°F/hr) From Various Intercritical Annealing Temperatures

Intercritical Annealing Temperature, °F	Yield Strength, ksi	Ultimate Tensile Strength, ksi	Total Elongation, %	Yield-Point Elongation, %	Yield Strength / Tensile Strength
		Steel A (1.5% Mn)			
1340	38.4	56.3	38.5	6.5	0.68
1380	39.0	54.4	36.0	3.4	0.72
1420	41.1	56.7	36.5	8.0	0.72
		Steel B (2.5% Mn)			
1300	31.3	74.9	29.0	0	0.42
1340	36.7	86.2	25.5	0	0.43
1380	41.9	73.4	27.0	0	0.57
1420	41.0	69.1	29.0	0	0.59
		Steel C (3.35% Mn)			
1300	42.4	108.6	20.0	0	0.39
1340	71.6	122.1	19.5	0	0.58
1380	82.3	116.9	17.0	0	0.70
1420	58.1	91.1	19.5	0	0.64

at 1340°F and then decreased as the annealing temperature was further increased to 1420°F. Also, like steel B, the yield/tensile-strength ratio of steel C was higher at the higher annealing temperatures (Figure 9).

To determine whether microstructural differences accounted for the observed changes in tensile properties as a function of annealing temperature, steel B (2.5% Mn) was examined by scanning electron microscopy. The change in microstructure with annealing temperature (and subsequent slow cooling) is shown in Figure 10. After annealing at 1300°F, a mixture of martensite and pearlite was present at the ferrite boundaries (Figure 10a). However, Figure 10b shows a grain-boundary phase consisting entirely of martensite after annealing at 1340°F. After annealing at 1380°F (Figure 10c), pearlite was again observed, together with some martensite, at the ferrite boundaries. As the temperature was further increased to 1420°F, Figure 10d shows that the proportion of martensite decreased further in favor of pearlite. Because annealing at 1340°F resulted in the maximum tensile strength and the maximum proportion of martensite for steel B, a dependence of tensile strength on the proportion of martensite is indicated. Furthermore, the observed increase in the yield/tensile-strength ratio for steels B and C (Figure 9) at temperatures above 1340°F probably resulted from the decreased proportion of martensite and associated increase in pearlite (Figures 10c and 10d).

(b) Fast Cooling (3000°F/hr)

Microstructures containing martensite were observed in steel A after fast cooling from all the annealing temperatures studied. Table V and Figure 11 show that the yield strength of steel A increased only slightly (from 29.5 to 33.3 ksi) as the annealing temperature was increased from 1340 to 1420°F. In contrast, the tensile strength remained relatively unchanged (at about 70 ksi) up to 1380°F and then decreased to 64.2 ksi at 1420°F. This finding is consistent with the observation of a maximum tensile

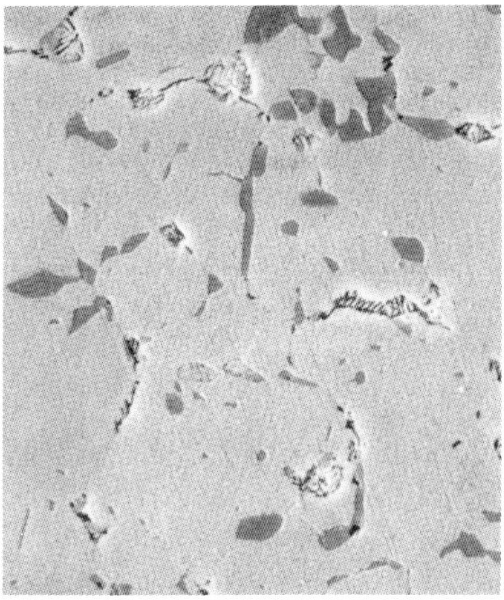

a) 1300°F

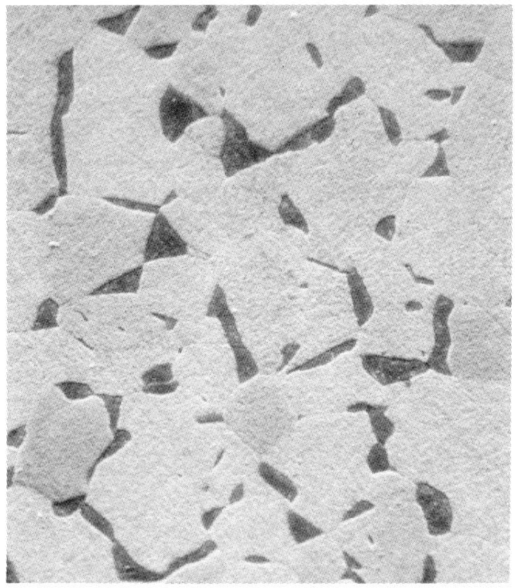

b) 1340°F

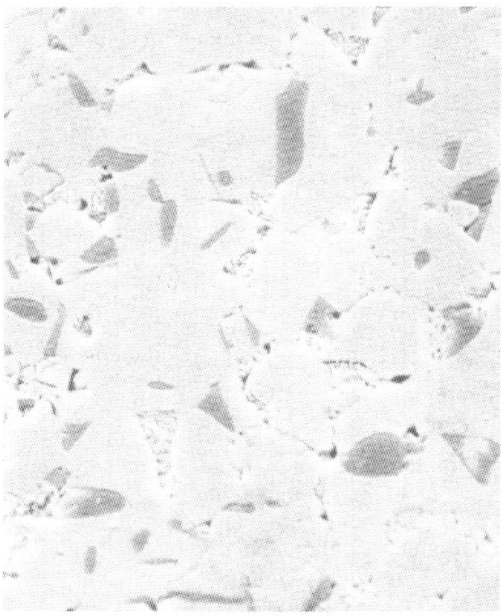

c) 1380°F

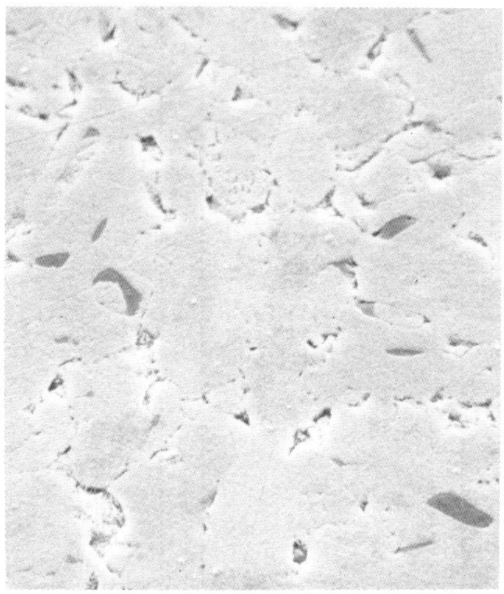

d) 1420°F

Fig. 10 - SEM micrographs of steel B (2.5% Mn) after annealing at various intercritical temperatures and slow cooling (25°F/hr). X3000.

Table V

Longitudinal Tensile Properties of Steels After Fast Cooling (3000°F/hr) From Various Intercritical Annealing Temperatures

Intercritical Annealing Temperature, °F	Yield Strength, ksi	Ultimate Tensile Strength, ksi	Total Elongation, %	Yield-Point Elongation, %	Yield Strength / Tensile Strength
		Steel A (1.5% Mn)			
1340	29.5	73.0	29.0	0	0.40
1380	32.2	67.8	28.0	0	0.50
1420	33.3	64.2	32.0	0	0.52
		Steel B (2.5% Mn)			
1300	37.4	91.5	22.0	0	0.41
1340	49.2	89.2	17.0	0	0.55
1380	61.2	91.6	22.0	0	0.67
1420	38.8	77.6	27.0	0	0.50
		Steel C (3.35% Mn)			
1300	59.0	120.5	18.5	0	0.49
1340	85.9	125.4	12.0	0	0.68
1380	77.4	105.6	15.0	0	0.73
1420	62.6	103.1	19.0	0	0.61

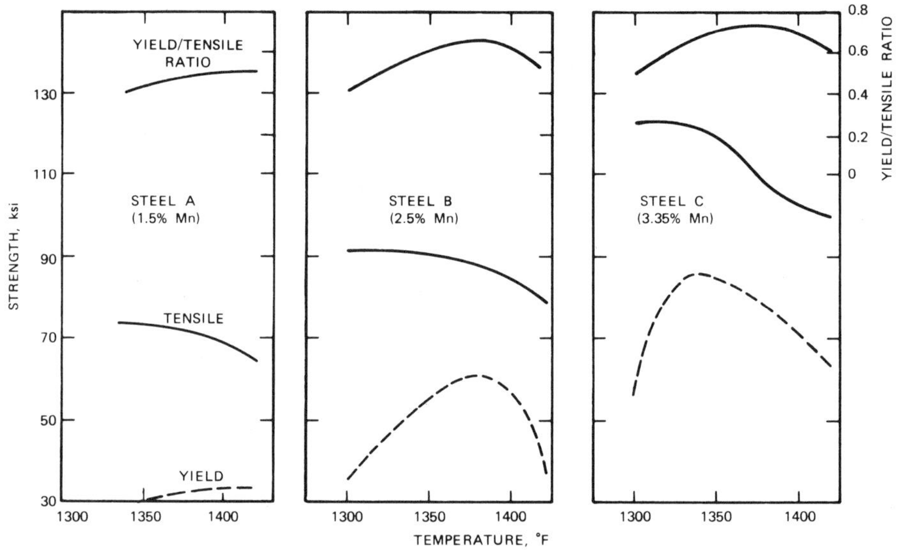

Fig. 11 - Tensile properties of steels as a function of intercritical annealing temperature and subsequent fast cooling (3000°F/hr).

strength at 1340°F for those slow-cooled steels that developed microstructures containing martensite. In addition, like the slow-cooled dual-phase steels (B and C), the yield/tensile-strength ratio of steel A increased from 0.40 to 0.52 as the annealing temperature increased from 1340 to 1420°F. For steels B and C that were fast-cooled from different annealing temperatures, the same trend in tensile properties was observed (Figure 11) as when the same steels were slow-cooled—that is, the yield and tensile strengths increased to a maximum at 1340 or 1380°F and then decreased at higher temperatures. For both slow and fast cooling (Figures 9 and 11, respectively), the yield/tensile-strength ratio increased as the annealing temperature increased. Although the yield/tensile-strength ratio usually decreased again at 1420°F, the ratio never decreased to that at 1300°F, where the lowest yield/tensile-strength ratio was consistently observed.

Dependence of Tensile Properties on Volume Fraction of Martensite

Because the findings discussed in the previous sections provided a strong indication that the tensile properties of the dual-phase steels depended on the microstructure, and especially on the proportion of martensite, a closer study was made of the correlation between tensile properties and the volume fraction of martensite. Scanning electron micrographs (such as those in Figures 2, 7, and 10) of all three steels were used for assessing the volume fraction of martensite by a quantitative-television-microscopy technique. In this technique, the dark martensite areas of the micrographs were traced onto transparent plastic sheets, and the volume fraction was assessed by using an epidiascope and a quantitative television microscope. Weak correlations were observed between the volume fraction of martensite and either the yield strength or the yield/tensile-strength ratio, whereas a good correlation with tensile strength was observed. In fact, Figure 12 shows a linear dependence of tensile strength on the volume fraction of martensite (which is shown as a range of the measured values).

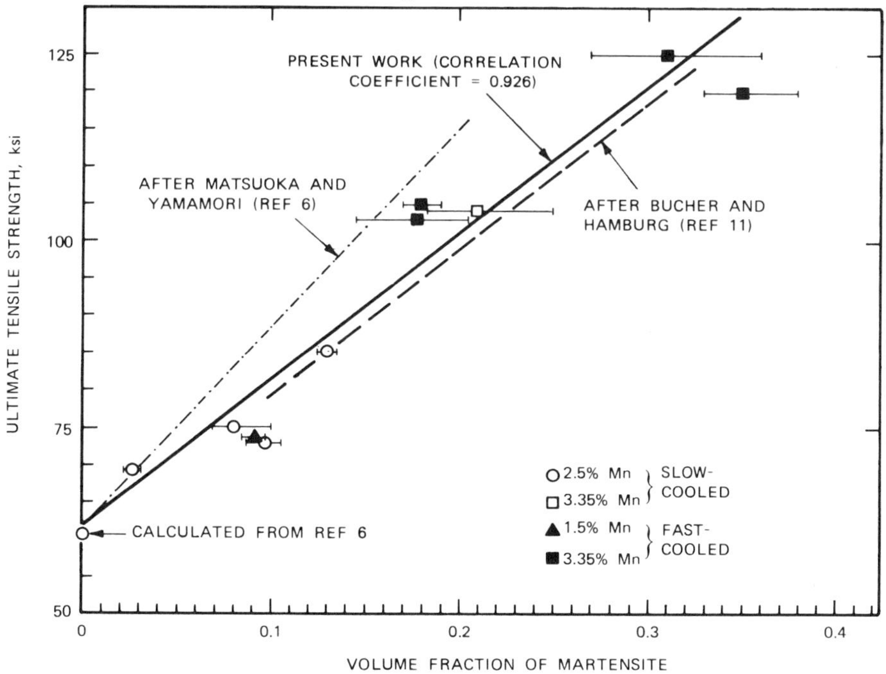

Fig. 12 - Dependence of tensile strength on volume fraction of martensite.

Surprisingly the data for steels A, B, and C, with different manganese contents, fit the same correlation line, and no adjustments were made for additional strengthening effects due to grain refinement or to solid-solution strengthening of the ferrite by manganese. Furthermore, the data obtained for the present steels agree closely with the correlation developed by Bucher and Hamburg[11] for air-cooled steels that contain about 1.5 percent manganese, and with the correlation obtained by Matsuoka and Yamamori.[6]

Thus, Figure 12 shows that the volume fraction of martensite largely determines the tensile strength of dual-phase steels. The steel composition and cooling rate from the annealing temperature do not appear to directly influence the tensile strength; rather, these variables indirectly influence the tensile strength by determining the amount of martensite formed. The relationship shown in Figure 12 indicates that a volume fraction of martensite of about 0.2 is needed to obtain a tensile strength of about 100 ksi. The observed dependence of tensile strength on the volume fraction of martensite is anticipated in view of the high work-hardening behavior of steels consisting of an aggregate of relatively soft ferrite and very hard martensite particles.[2] However, the yielding of martensite-containing dual-phase steels is dependent on the nature of the ferrite matrix, which in turn depends on several factors, such as the presence of mobile dislocations, the presence of residual stresses in the ferrite, the manganese content and the grain size of the ferrite.

To explain the influence of the cooling rate and of the intercritical annealing temperature on the amount of martensite formed in dual-phase steels, and hence on their tensile strengths, it is necessary to consider the transformation of the carbon-rich austenite that is formed at the intercritical annealing temperature. Because manganese decreases the eutectoid composition of steels,[9] the carbon content of austenite in equilibrium with ferrite at low intercritical temperatures (1300 to 1340°F) decreases from about 0.6 to 0.45 percent as the manganese increases from 1.5 to 3.35 percent in steels A and C, respectively. Unfortunately, no continuous-cooling-transformation (CCT) diagrams could be found in the published literature for such high-carbon high-manganese steels.

However, for a given manganese content, fast cooling probably either avoids or just intersects the pearlite and bainite noses of the CCT diagram of the high-carbon material (about 0.40 to 0.6%), shown schematically in Figure 13a, resulting in a high proportion of martensite. By comparison, the slow-cooling curve intersects the CCT diagram well into the ferrite-pearlite area, so that a lower proportion of martensite forms and results in lower tensile strengths than those of the fast-cooled steel. When higher intercritical annealing temperatures are used, the carbon content of the austenite is decreased to intermediate carbon levels (about 0.2 to 0.3%), and the corresponding CCT diagram is moved to the left, as shown in Figure 13b, so that the steel exhibits lesser hardenability. Thus, for a given cooling rate, less martensite forms after annealing at the higher intercritical temperature than at a low temperature. This explanation is confirmed by the metallographic study for steel B (2.5% Mn) shown in Figure 10.

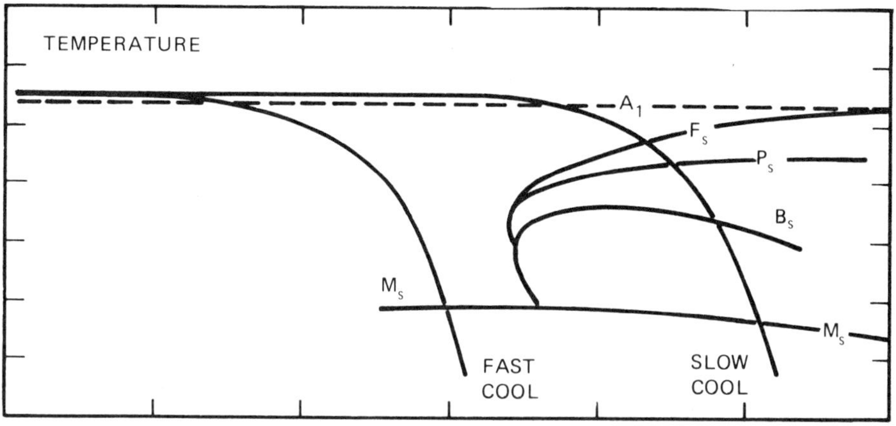

(A) LOW INTERCRITICAL ANNEALING TEMPERATURE; AUSTENITE HAS HIGH CARBON CONTENT

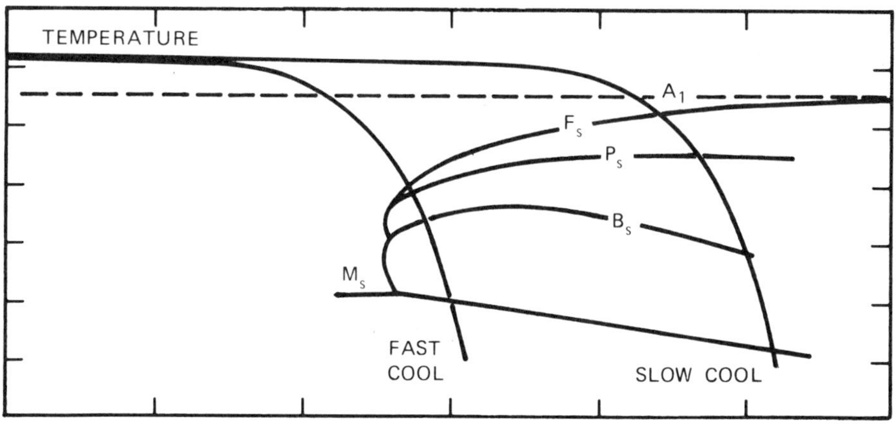

(B) HIGH INTERCRITICAL ANNEALING TEMPERATURE; AUSTENITE HAS INTERMEDIATE CARBON CONTENT

Fig. 13 - Schematic continuous-cooling diagrams for austenite formed at different intercritical annealing temperatures.

Summary and Conclusions

The current work supports earlier findings of Matsuoka and Yamamori that dual-phase microstructures (ferrite and martensite) can be obtained for sheet steels containing at least 2.5 percent manganese when they are slowly cooled (as when box-annealed and cooled at about 25°F/hr) from intercritical annealing temperatures in the range 1300 to 1420°F. By increasing the manganese content to about 3.35 percent, the tensile strength of the slowly cooled dual-phase steel was increased from 91 ksi to about 108 ksi, whereas the yield strength was increased to a lesser extent (from 31 to 43 ksi). For steels containing manganese contents of 1.5 percent, slow cooling from intercritical annealing temperatures failed to develop the dual-phase microstructures and the characteristic low yield strengths and high tensile strengths. However, for faster cooling rates (3000°F/hour) that approach air cooling of sheet, dual-phase microstructures were obtained for 1.5 percent manganese steels. After faster cooling from a given intercritical temperature, the yield strengths of the dual-phase steels containing 2.5 to 3.35 percent manganese were increased by at least 7 and 12 ksi, respectively. Thus fast cooling rates can be used to develop dual-phase structures in lower-manganese (1.5%) steels and/or to increase the yield and tensile strengths of the high-manganese (2.5 to 3.5%) dual-phase steels.

The presence of martensite, pearlite, and ferrite in the slow-cooled steels, and of martensite, very fine pearlite, or bainite and ferrite in the fast-cooled steels, was confirmed by scanning and transmission electron microscopy. A high density of dislocations was observed in the ferrite adjacent to the martensite areas. These dislocations, together with the likely presence of residual stresses in the ferrite, are believed to account for the continuous-yielding behaviors (absence of a yield-point elongation), the low yield strengths, and the low ratios of yield to tensile strength observed for the dual-phase steels. From quantitative metallographic studies it was found that the tensile strength of the dual-phase steels was dependent on the volume fraction of martensite (f_m); an f_m value of about 0.2 appears to be necessary to produce a tensile strength of about 100 ksi.

For a given manganese content and a given cooling rate, intercritical annealing at low temperatures (1300 to 1340°F) was found to result in low ratios of yield to tensile strength desired for good formability in sheet forming operations.

For the most part, intercritical annealing at high temperatures (such as 1420°F) and then cooling to form dual-phase structures caused the tensile strength of the steel to decrease. This effect was found to be associated with the formation of lower volume fractions of martensite. The reduced amount of martensite probably results from the formation, and subsequent transformation, of a lower carbon austenite formed at the higher annealing temperature.

Acknowledgments

The authors are indebted to F. C. Bell for developing the procedures for distinguishing the various microstructural constituents by scanning electron microscopy and to D. A. Gomrick and B. J. Sobek for assistance with the transmission electron microscopes. The authors are also grateful to G. R. Speich and R. L. Miller for useful discussions.

References

1. R. A. Grange, "Fibrous Microstructures Developed in Steel by Thermomechanical Processing," Proceedings of Second International Conference on Strength of Metals and Alloys, Volume 111, 1970, pp. 861-876.

2. D. S. Dabkowski and G. R. Speich, "Transformation Products and Stress-Strain Behavior of Controlled Rolled Mn-Mo-Cb Line Pipe Steels," Mechanical Working and Steel Processing XV, AIME, 1977, pp. 284-312.

3. G. Tither and M. Lavite, "Beneficial Stress-Strain Behavior of Moly-Columbium Steel Line Pipe," Journal of Metals, September 1975, pp. 15-23.

4. M. S. Rashid, "GM 980X—Potential Applications and Review," presented at International Automotive Engineering Congress, Detroit, Michigan, February 28 to March 4, 1977, SAE preprint No. 770211.

5. D. J. Bailey, "A Family of Thermomechanically Treated High-Strength Sheet Steels" presented at the SAE Automobile Engineering Meeting, Dearborn, Michigan, October 18 to 22, 1976, SAE preprint No. 760715.

6. T. Matsuoka and K. Yamamori, "Metallurgical Aspects in Cold-Rolled High-Strength Steel Sheets," Metallurgical Transactions, Volume 6A, 1975, pp. 1613-1622.

7. S. Hayami and T. Furukawa, "A Family of High-Strength, Cold-Rolled Steels," Proceedings of Micro-Alloying 75 (International Symposium on High-Strength Low-Alloy Steels), Washington, D. C., October 1 to 3, pp. 311 to 320.

8. R. A. Grange, "Estimating Critical Ranges in Heat Treatment of Steels," Metal Progress, April 1961, pp. 73 to 75.

9. E. C. Bain and H. W. Paxton, Alloying Elements in Steel, 2nd ed., American Society for Metals, 1961, p. 112.

10. G. R. Speich and W. C. Leslie, "Tempering of Steel," Metallurgical Transactions, Volume 3, 1972, pp. 1043-1054.

11. J. H. Bucher and E. G. Hamburg, "High Strength Formable Sheet Steel," presented at the International Automotive Engineering Congress, Detroit, Michigan, February 28 to March 4, 1977, SAE preprint No. 770164.

DISCUSSION

T. E. Parker, The Steel Company of Canada Ltd., Hamilton, Ontario

1. Do you anticipate any problems in weldability (both resistance spot and flash butt) for these very high manganese batch annealed grades?

2. In view of the high manganese requirements for successfully producing dual-phase by batch annealing, do you anticipate the need for special steelmaking practices to maintain low carbon contents typical of contemporary formable and weldable HSLA grades?

Authors' Reply

1. Small laboratory melted heats were used in this investigation and therefore, extensive welding studies could not be conducted because of the limited quantity of material. However, several peel tests were made on spot welded samples and full nugget pullout resulted indicating that the welds were ductile. Therefore, it is anticipated that such steels can be successfully spot welded.

   We have no data on which to predict the flash butt weldability of the high manganese steels.

2. To date we have not melted any production heats of the high manganese steels, and therefore we have no firsthand experience regarding steelmaking practices for such steels. However we believe it should be possible to make the high-manganese steels using existing steelmaking practices with some modified procedures.

DISCUSSION

I. Gupta, Inland Steel Research Laboratories, East Chicago, Indiana

Do you think that the batch-annealed approach will be commercially feasible?

Authors' Reply

The work presented was only for laboratory heats — no production trial data is available as yet. However, we are optimistic that the controlled-rolled batch-annealed (CRBA) dual-phase steels can be produced on production facilities. Temperature variations can be controlled and the temperature sensitivity of the properties is not that severe. Our optimism is supported by the work of Matsuoka and colleagues at Sumitomo Metal Industries, Ltd., Japan who have run extensive production trials involving CRBA dual-phase steels containing 2.5 Mn.

DEVELOPMENT AND MILL TRIAL OF AS-ROLLED DUAL-PHASE STEEL

A. Phillip Coldren
Geoffrey Tither

Climax Molybdenum Co. of Michigan
a Subsidiary of AMAX Inc.
Ann Arbor, Michigan

Arden Cornford
John R. Hiam

Dominion Foundries and Steel, Ltd.
Hamilton, Ontario
Canada

Abstract

Dual-phase steels are high-strength steels with the formability of conventional steels of much lower strength, the properties having been associated with a two-phase structure of ferrite and martensite. To date, the development of these alloys has concentrated on the use of heat-treatment equipment, i.e., continuous annealing and normalizing lines, for producing the dual-phase structure. Since the availability of such facilities is very limited and because the cost of such annealing is high, the development of as-rolled dual-phase steels would be commercially desirable.

This paper deals with the technical principles involved in the development of an as-rolled dual-phase steel containing 0.065% C max, 1.2% Mn, 0.9% Si, 0.6% Cr and 0.4% Mo. The results of laboratory experiments and a commercial plant trial are given which show that the concept has considerable promise.

Introduction

The weight of an automobile is one of the most important factors affecting its average rate of fuel consumption. Increasingly stringent federal regulations are forcing automobile manufacturers to increase the fuel efficiency of their cars, and a major part of the current improvement is being achieved by the most direct approach to weight reduction--size reduction. Since this trend toward smaller cars is expected to meet with increasing consumer resistance, future auto designs will no doubt be based more and more on increased strength-to-weight ratios of component parts and of the overall structure of the car. This will require greater use of high strength steels, aluminum, plastics and composites.

The use of high strength low alloy (HSLA) steels for automotive applications has in the past been limited by their relatively poor formability. However, the current development of the new dual-phase steels[1-3]--highly formable HSLA steels with 655 MPa (95 ksi) tensile strength--is beginning to provide automotive engineers with a lower cost alternative to the use of aluminum for reducing the weight of automobiles. Dual-phase steels exhibit their high ductility and formability with a microstructure of martensite (plus austenite) islands in a matrix of polygonal ferrite. The reason for their superior formability relative to conventional HSLA steels remains to be established, but most investigators agree that a high degree of ductility in the ferrite matrix and the absence of cracking within the martensite and at the martensite-ferrite interface is necessary, if not sufficient, to produce the superior properties.

Methods of producing dual-phase steels have until now involved heat treatment of a previously hot-rolled or cold-rolled strip steel. The heat treatment included either intercritical or slightly supercritical heating followed by controlled cooling to produce the ferrite-martensite structure. The present paper deals with the development of a steel that can be produced with the desired dual-phase structure in the as-hot-rolled condition. It will be of interest to those steel companies desiring to produce dual-phase steels in hot-roll gauges, but who do not have ready access to the necessary continuous heat treatment facilities.

The first part of the paper contains the results of experiments performed at the Climax Molybdenum Laboratory where the Mn-Si-Cr-Mo as-rolled dual-phase steel was developed, and the second part describes the first mill trial carried out by Dominion Foundries and Steel, Ltd. (Dofasco) of Hamilton, Ontario, in August 1977.

Climax Molybdenum Laboratory Program

Experimental Approach

The objective of the alloy development program was to find a steel composition that would respond in the following way to normal processing on a standard hot-strip mill: (a) transform to 80 to 90% polygonal ferrite after the final roll pass and before entering the coiler, (b) avoid transformation of the remaining 10 to 20% austenite (carbon enriched) to bainite, and (c) transform to 10 to 20% martensite much later during slow cooling in the coil. The fact that martensite islands can be produced in a hot-mill coiled product was known from previous experience with heavy gauge Mn-Mo-Nb steel for line pipe. It had also been observed that the alloying elements manganese, silicon, chromium and molybdenum promoted the formation of martensite islands. The challenge was to find a combination of composition and processing that would produce both a large amount of polygonal ferrite and the martensite islands in the same coiled steel strip. The solution was conceived to be (a) a balance between the ferrite-promoting tendencies of austenite deformation accumulated after cessation of recrystallization, low carbon content and high silicon content on the one hand, and the pearlite-delaying effects of molybdenum, manganese and chromium on the other, and (b) a stabilization of the carbon-enriched islands of austenite by all the elements to prevent immediate transformation to bainite in the coiling temperature range [e.g., 510 to 620°C (950 to 1150°F)] and thereby allow martensite to form later at a lower temperature.

An exploratory program was initiated to find an appropriate composition, using as guides existing compilations of continuous cooling transformation (CCT) diagrams[4,5] and past alloy development experience at the Climax Molybdenum Laboratory. The program involved a basic cycle of melting, forging, rolling and metallographic screening, followed by tensile testing of those steels that exhibited dual-phase or near-dual-phase structures. The cycle was then repeated with new compositions selected on the basis of the results of the previous cycle.

Laboratory Procedures

Melting and Forging. The experimental steels were induction melted in an argon atmosphere to facilitate nitrogen control, and they were aluminum killed. The heat size was 34 kg (75 lb), and two of the heats were each split cast as four ingots to allow variations to be made in the carbon content and in the phosphorus content. Buttons cast as an integral part of each ingot were used for chemical analysis. The compositions of the laboratory steels described in the present paper are given in Table I.

Table I. Laboratory Steel Compositions

Steel No.	Climax Heat No.	Element, Wt. %								
		C	Mn	Si	Mo	Cr	Al	N	P	S
1	P1626B[a]	0.065	1.19	0.87	0.38	0.61	0.06	0.005	0.010	0.005
2	P1697	0.068	1.20	0.91	0.38	0.60	0.04	0.005	0.014	0.008
3	P1724A	0.052	1.36	0.88	0.38	0.60	0.02	0.003	0.003	0.006
4	P1724B	0.053	1.34	0.88	0.38	0.60	(0.02)[b]	0.002	0.006	0.006
5	P1724C	0.055	1.32	0.86	0.37	0.60	(0.02)	0.002	0.010	0.006
6	P1724D	0.054	1.33	0.91	0.38	0.61	(0.02)	0.002	0.026	0.006
7	P1796B	0.085	0.83	0.85	0.39	0.60	0.05	0.004	0.008	0.004
8	P1720B	0.092	0.81	0.90	0.39	0.58	0.06	0.005	0.014	0.006
9	P1805	0.11	0.82	0.85	0.40	0.72	0.04	0.003	0.005	0.004
10	P1673	0.032	1.21	0.92	0.39	0.60	0.03	0.005	0.013	0.004
11	P1674	0.075	1.20	0.92	0.39	0.60	0.03	0.005	0.013	0.005
12	P1675	0.091	1.21	0.92	0.46	0.60	0.02	0.005	0.013	0.004
13	P1676	0.12	1.20	0.91	0.39	0.60	0.02	0.004	0.013	0.004

[a] A letter at the end of a heat number indicates the sequence of the ingot in a split heat.

[b] Parentheses = assumed value based on analysis of the first ingot in the split heat.

The ingots were press-forged to 25-mm (1-in.) thick slabs and cut to 25-mm by 51-mm by 127-mm (1-in. by 2-in. by 5-in.) rolling blanks.

Simulation of Hot Strip Mill Processing. The schematic representation in Figure 1 provides details of the laboratory simulation of hot strip mill processing of 2.5-mm (0.10-in.) gauge coiled sheet, showing the relationship of the temperature-time profile to the CCT diagram. After one hour in the reheat furnace at 1260°C (2300°F), a 25-mm (1-in.) thick slab was descaled and rolled with a six-pass schedule to 2.5-mm (0.10-in.) gauge sheet in 50 to 40 seconds (total time out of furnace), depending on the desired finish-rolling tmeperature [760 to 925°C (1400 to 1700°F)]. Temperatures during rolling were monitored with an IRCON* infrared pyrometer on the entry side of the mill. Five seoncds after completion of the last pass, the sheet was clamped between water cooled steel plates and accelerated cooled at approximately 28°C/sec (50°F/sec) to the simulated coiling temperature. The five-second delay and the duration of the platen cooling were both automatically controlled by timers in the circuit that actuated a pneumatic cylinder to lower and raise the upper steel plate. Coil cooling was simulated in a re-circulating-air furnace that was program controlled to cool at a rate of 28°C/hr (50°F/hr). The sheets were removed from the furnace at about 100 to 150°C (210 to 300°F).

Tensile Testing. Tensile testing was performed on duplicate longitudinal specimens having a reduced section of 57 mm by 13 mm by 2.5 mm (2.25 in. by 0.5 in. by 0.1 in.). Strain rates of 18 and 450%/hr were employed in the 0-to-3% and the 3%-to-fracture strain ranges, respectively. Load elongation curves were recorded to fracture, and uniform elongation values were taken from these curves. Total elongation was determined from punch marks on the fractured specimens.

---

*Trade name

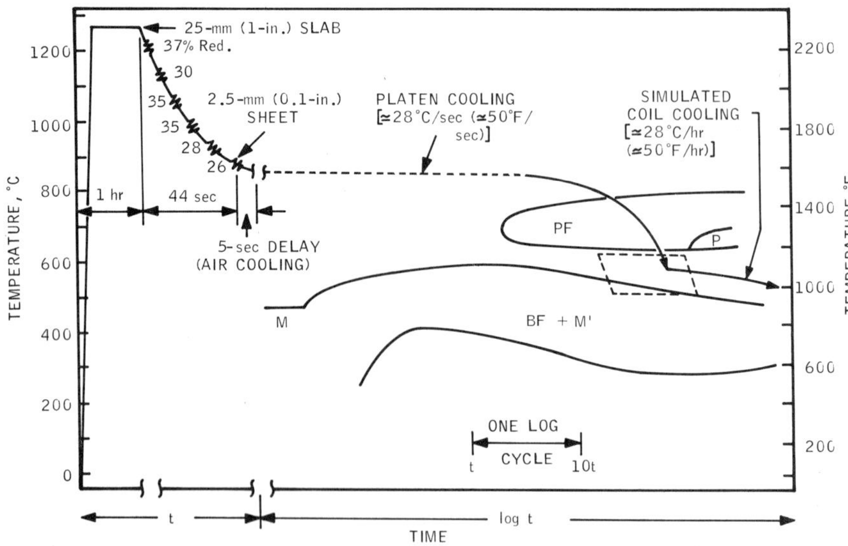

Figure 1 Schematic Representation of Laboratory Simulation of Hot-Strip Mill Processing

Metallography. Conventional optical microscopy was used for screening out those sheets that did not exhibit the dual-phase structure.

Scanning electron microscopy was employed to measure ferrite grain size and volume percent of the microconstituents. The specimens were final polished on a Syntron vibratory polisher, and then etched approximately 20 seconds in an etchant consisting (by weight) of 4% picric acid plus 1% nitric acid in ethanol. A gold-palladium alloy coating 20 nm thick was applied to the specimens by vapor deposition to prevent charge buildup in the electron beam. The grain size values were determined by the linear intercept method, using a total traverse distance of 700 mm (28 in.) on each of five micrographs taken at 2000X. Volume percent determinations of the martensite-austenite and the bainitic ferrite constituents were made on selected specimens by point counting on the same five micrographs, using a 285-point grid. The five micrographs were from the mid-thickness, quarter-thickness and near-surface positions of longitudinal sections. The use of a high magnification was necessitated by the extreme fineness of the structures.

X-Ray Determination of Retained Austenite. Retained austenite contents were measured on selected specimens using an X-ray diffraction technique in which a polycrystalline specimen was rotated and oscillated (tilted) in a beam of Mo $K\alpha$ radiation. The integrated X-ray intensities were determined for the principal alpha and gamma peaks, and the volume percent retained austenite was then calculated from the integrated intensities. The plane of examination was parallel to the rolling direction and perpendicular to the rolling plane. The values obtained were accurate to within plus or minus one volume percent or less of retained austenite.

Continuous Cooling Transformation Diagram. A continuous cooling transformation (CCT) diagram was determined for a steel containing 0.065% C, 1.19% Mn, 0.87% Si, 0.61% Cr and 0.38% Mo (Steel No. 1). A dilatometric method was employed, and the equipment used was a Formastor-F dilatometer which employs induction heating *in vacuo* of a 3-mm diameter by 10-mm long cylindrical specimen. The specimens were austenitized 20 minutes at 960°C (1760°F), 30°C (54°F) above the Ac_3 temperature determined for that steel. Cooling from the austenitizing temperature was controlled either by a programmed reduction in power input or by helium quenching. The dilatometric data were supplemented by metallographic examinations of the specimens.

Experimental Results

More than 60 laboratory steels were processed and examined in this program. The thirteen steels that are described in this paper (Table I) were chosen because they demonstrated (a) the best properties obtained and (b) the effects of carbon and phosphorus on the strength-ductility balance. The detailed tensile test results and the grain size data for these steels are summarized in Table II.

Strength-Ductility Balance. A low 0.2% offset yield strength may be important in some forming operations, but it is generally agreed that the most significant and useful characteristic of dual-phase steels is that they exhibit unprecedented combinations of high ductility with high tensile strength which, combined with high values of uniform elongation (strain at the onset of necking), lead to excellent formability. The strength-ductility properties of two groups of as-rolled dual-phase steels were determined. In one group the manganese-to-carbon ratio was high--in the range of 18 to 26. This group of steels showed the most promise in the laboratory investigations. Steels with lower Mn/C ratios--ranging from 8 to 10--were also investigated because commercially they would be more economical to produce. Both groups contained nominally 0.9% Si, 0.6% Cr and 0.4% Mo.

Table II Summary of Tensile Test Results[a] and Grain Size Data for Laboratory Steels

Steel No.	Finish-Rolling Temp., °C (°F)	Simulated Coiling Temp., °C (°F)	Ferrite Grain Size, ASTM No.	Offset Yield Strength, MPa (ksi)			Plastic Strain at 551 MPa (80 ksi), %	Tensile Strength, MPa (ksi)	Elongation, % in 51 mm (2 in.)	
				0.2%	2.0%	3.0%			Uniform[b]	Total
1[c]	760 (1400)[d]	620 (1150)	13.4	396 (57.5)	522 (75.8)	553 (80.2)	3.0	647 (93.9)	15	24.0
2	870 (1600)	620 (1150)	13.2	361 (52.4)	501 (72.7)	535 (77.7)	3.4	639 (92.7)	16	24.2
3	760 (1400)	615 (1140)	13.7	442 (64.1)	574 (83.3)	606 (87.9)	1.5	674 (97.8)	10	23.0
		600 (1110)	14.0	448 (65.0)	570 (82.8)	600 (87.1)	1.5	679 (98.5)	13	24.7
		570 (1060)	14.1	382 (55.4)	535 (77.6)	569 (82.6)	2.5	664 (96.3)	15	23.8
4	760 (1400)	615 (1140)	13.5	416 (60.4)	531 (77.0)	570 (82.7)	2.5	657 (95.4)	15	24.4
		600 (1110)	13.8	420 (60.9)	563 (81.7)	594 (86.2)	1.7	688 (99.8)	14	22.9
		570 (1060)	13.4	389 (56.5)	544 (79.0)	580 (84.2)	2.3	693 (100.6)	14	25.1
5	760 (1400)	615 (1140)	13.6	431 (62.5)	566 (82.2)	598 (86.8)	1.7	676 (98.1)	12	21.6
		570 (1060)	13.3	463 (67.2)	568 (82.4)	589 (85.5)	1.5	673 (97.7)	12	19.4
6	760 (1400)	615 (1140)	13.8	485 (70.4)	626 (90.8)	670 (97.2)	0.6	698 (101.3)	--	19.6
		570 (1060)	13.4	476 (69.1)	528 (76.7)	553 (80.3)	3.0	637 (92.5)	15	23.6
7	925 (1700)	615 (1140)	ND[e]	430 (62.4)	467 (67.8)	501 (72.7)	5.0	611 (88.7)	18	26.0
		615 (1140)	ND	433 (62.9)	459 (66.6)	488 (70.8)	5.4	611 (88.7)	18	26.1
		605 (1120)	ND	389 (56.5)	474 (68.8)	511 (74.1)	4.4	633 (91.9)	16	25.6
		605 (1120)	ND	453 (65.7)	524 (76.0)	553 (80.2)	3.0	643 (93.3)	13	22.4
8	870 (1600)	610 (1130)	ND	389 (56.5)	489 (71.0)	524 (76.1)	4.2	633 (91.8)	16	25.5
		595 (1100)	ND	460 (66.7)	513 (74.5)	546 (79.3)	3.2	654 (95.0)	14	25.3
		575 (1070)	ND	389 (56.5)	555 (80.6)	590 (85.6)	2.0	677 (98.3)	14	22.0
9	925 (1700)	615 (1140)	ND	579 (84.0)	619 (89.9)	647 (93.9)	--	702 (101.9)	8	16.6
		605 (1120)	ND	596 (86.5)	697 (101.2)	715 (103.8)	--	755 (109.6)	6	14.4
		590 (1090)	ND	597 (86.7)	692 (100.4)	715 (103.8)	--	766 (111.2)	9	16.4
	870 (1600)	615 (1140)	ND	567 (82.3)	599 (87.0)	630 (91.4)	--	719 (104.3)	10	19.2
		605 (1120)	ND	564 (81.8)	624 (90.6)	650 (94.4)	--	719 (104.3)	9	18.7
		590 (1090)	ND	506 (73.4)	601 (87.3)	634 (92.0)	0.4	730 (106.0)	9	18.7
	925 (1500)	615 (1140)	ND	580 (84.2)	597 (86.7)	622 (90.3)	--	693 (100.6)	10	20.8
		605 (1120)	ND	613 (89.0)	684 (99.3)	693 (100.6)	--	763 (110.8)	9	18.3
		590 (1090)	ND	601 (87.2)	664 (96.3)	688 (99.9)	--	745 (108.2)	8	17.1
10	760 (1400)	620 (1150)	13.1	431 (62.6)	506 (73.5)	535 (77.6)	3.6	611 (88.7)	14	23.0
		595 (1100)	12.6	434 (61.6)	524 (76.1)	553 (80.3)	2.9	629 (91.3)	14	20.0
11	760 (1400)	620 (1150)	13.4	440 (63.8)	553 (80.3)	609 (88.4)	2.0	683 (99.1)	11	18.0
		595 (1100)	13.8	548 (79.6)	659 (95.6)	689 (100.0)	0.2	757 (109.8)	11	18.0
12	760 (1400)	595 (1100)	14.3	524 (76.1)	686 (99.6)	729 (105.5)	0.3	804 (116.7)	11	18.0
		565 (1050)	15.8	714 (103.6)	808 (117.2)	841 (122.1)	--	887 (128.7)	6	11.0
13	760 (1400)	595 (1100)	14.4	469 (68.1)	710 (103.1)	767 (110.7)	0.5	861 (125.0)	10	16.0
		565 (1050)	16.1	508 (73.8)	758 (110.0)	803 (116.6)	0.4	871 (126.4)	7	12.0

[a] Average values from duplicate tests on longitudinal specimens. Individual test values from selected steels are plotted in Figures 2 and 3.

[b] From load-elongation curve.

[c] See Table 1 for composition.

[d] Note: It was determined later that 870°C (1600°F) is preferable to 760°C (1400°F) as a finish-rolling temperature.

[e] ND = Not Determined.

Individual test results for the high Mn/C ratio steels (Steels 1 through 6) are presented in Figure 2 on a plot of total elongation versus tensile strength. For comparison, Figure 2 also contains the strength-ductility relationship for conventional microalloyed HSLA steels (solid line), and the scatter band (dashed lines) published by Bucher and Hamburg[2] for VAN-QN, the best known heat-treated dual-phase steel commercially available. Three different data point symbols were used to show whether any trends exist with respect to variations in the coiling temperature range.

The most important conclusion drawn from Figure 2 was that in the tensile strength range of approximately 635 to 695 MPa (92 to 101 ksi), the ductility of the laboratory as-rolled dual-phase steels was superior to that of conventional HSLA steels and in many instances equivalent to that of VAN-QN. It can

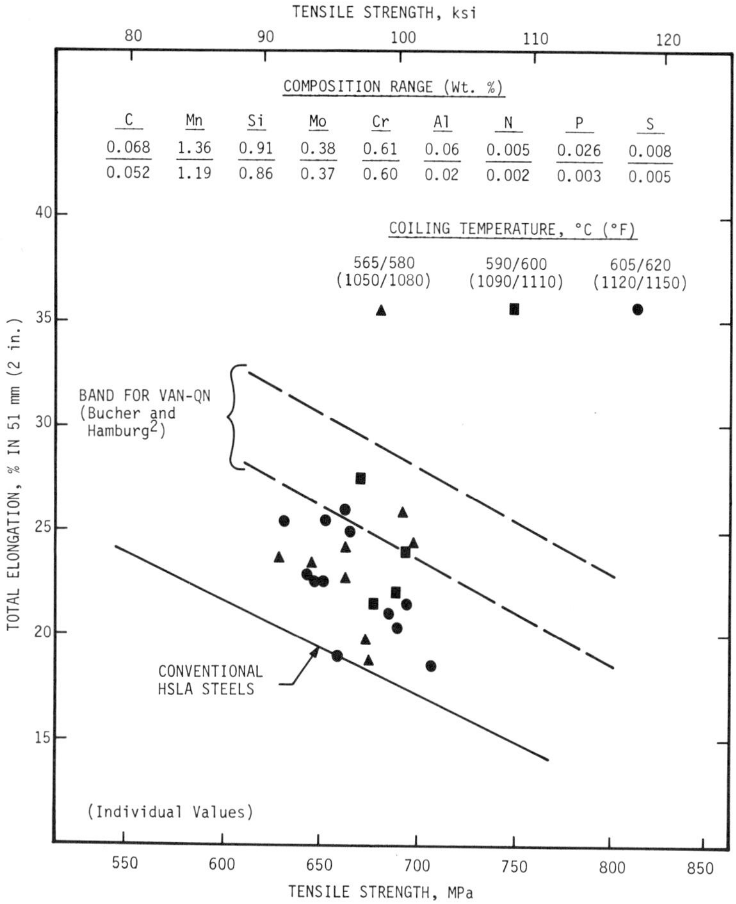

Figure 2 Strength-Ductility Balance of High Mn/C Ratio Laboratory Steels 1-6 Compared with That of Heat-Treated Dual-Phase Steel (VAN-QN) and Conventional Microalloyed HSLA Steels

also be seen that the circles, squares and triangles were randomly mixed, indicating that the properties were not affected by variations of coiling temperature in the range of 565 to 620°C (1050 to 1150°F).

The use of 760°C (1400°F) as a finish-rolling temperature was later found to be detrimental to the total elongation for some steels. Specifically, those steels having an Ar_3 above the finishing temperature and whose final structure therefore contained some deformed ferrite exhibited the lowest values of elongation seen in Figure 2.

Individual test results for the low Mn/C ratio steels (Steels 7, 8 and 9) are plotted in Figure 3, with the 0.09% C data points being distinguished from the 0.11% C data points. The band for VAN-QN is reproduced as a frame of reference.

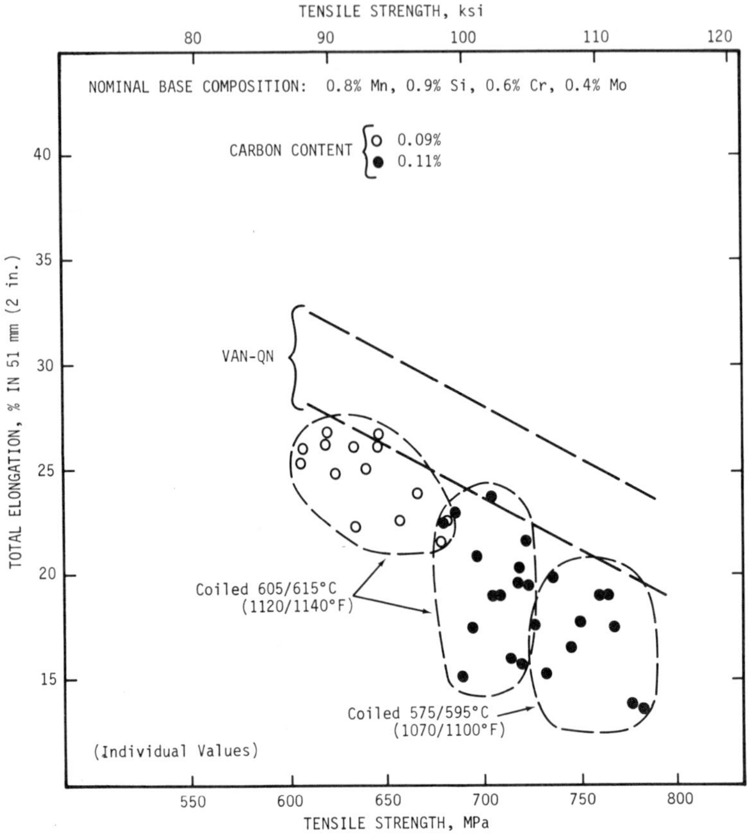

Figure 3 Effect of Carbon Content and Coiling Temperature on the Strength-Ductility Balance of Low Mn/C Ratio Laboratory Steels 7, 8 and 9

The low Mn/C ratio steels exhibited two characteristics that would create serious problems in commercial production. The first problem was that the properties were sensitive to small variations in carbon content. The steel represented by the solid points was only two points (0.02%) higher in carbon content than the steels represented by the open points, but the higher carbon steel exhibited significantly higher tensile strength and markedly lower elongation values. The other problem was a high sensitivity to small changes in coiling temperature. A 22°C (40°F) change in coiling temperature produced a 55 MPa (8 ksi) change in average tensile strength and a three percentage point change in the average total elongation.

Effect of Carbon in Steels with High Mn/C Ratio. Since the high Mn/C ratio steels appeared to be better suited for commercial production (good strength-ductility balance and low sensitivity to coiling temperature variations), it was decided to investigate the effect of varying carbon content within this base. For that purpose, Steels 10 through 13 were prepared, rolled to 2.5-mm (0.1-in.) gauge sheet and tested. Combined with the data from Steel 1, the new data provided a five-point carbon series from 0.032% C to 0.12% C.

The volume percent values of the microconstituents were determined for this series of steels. They are tabulated in Table III, and in Figure 4 they are plotted for a simulated coiling temperature of 595°C (1100°F). Between approximately 0.03% and 0.06% carbon, the amount of the M-A constituent increased slowly from approximately 6% to 12%. At about 0.065% carbon, the amount of bainitic ferrite suddenly increased at a rapid rate, reaching 33% (plus 27% M-A constituent) at 0.12% carbon. This was interpreted to mean that the increasing carbon content had shifted the polygonal ferrite nose to the right on the CCT diagram, decreasing the amount of polygonal ferrite that could form at the cooling rate employed and thereby allowing increasing amounts of bainitic ferrite to form at the lower temperatures. The amount of M-A constituent continued to increase moderately at the higher carbon contents, primarily as M-A islands within the bainitic ferrite regions.

The changing microstructures were reflected in the tensile properties as shown indirectly in Figure 5 and more directly in Figure 6. Initially the 0.2% offset yield strength decreased as the amount of the M-A constituent increased. This decrease in yield strength was due to a lowering of the proportional limit by the mobile dislocations and the residual stresses introduced into the ferrite matrix as a result of the shape change and volume expansion associated with the martensite transformation. With the nominal base composition of 1.2% Mn, 0.9% Si, 0.6% Cr, 0.4% Mo, the 0.2% yield strength appeared to reach a minimum value at about 0.05% carbon.

When bainitic ferrite in an amount as low as 10% was introduced, the 0.2% yield strength increased very abruptly by about 160 MPa (23 ksi), as shown in Figure 5. The tensile strength and 3% yield strength values also increased rather abruptly, but by smaller amounts. With further increases in carbon content the M-A and bainitic ferrite contents continued to increase and to raise the tensile strength and 3% yield strength, though, in this case, at decreasing rates. The 0.2% yield strength, however, decreased again as the amount of martensite increased.

In view of the strong effect of carbon above 0.065% in restricting the formation of polygonal ferrite, a recommended maximum carbon content of 0.065% was set, and a preferred carbon content of 0.05% was established on the basis of the minimum in the 0.2% yield strength curve observed at that carbon level.

Tensile and yield strength data from sheets simulated coiled at 565, 595 and 620°C (1050, 1100 and 1150°F) were plotted against volume percent M-A constituent to determine whether there were basic correlations independent of coiling temperature. As can be seen in Figure 6, the tensile strength exhibited a linear correlation, but the 0.2% yield strength data followed a complex curve similar to that in Figure 5. The linear correlation with tensile strength indicated that a simple rule of mixtures (ferrite plus martensite) governed that property. Apparently the presence of bainitic ferrite with the M-A constituent did not affect the correlation with tensile strength, but it produced a complex relationship with yield strength. Even when the two strength parameters were plotted against the sum of M-A constituent plus bainitic ferrite contents (not shown), no correlation was obtained with yield strength. The tensile strength data formed a curved-line rather than straight-line relationship.

The ferrite grain size, indicated by each point on the yield strength curve in Figure 6, showed no correlation with the yield strength values. Regarding tensile strength, it can be seen that the tensile strength correlation line would have a lower slope if each point were corrected to one common grain size number.

Table III Effect of Carbon and Coiling Temperature on Microstructure of 2.5-mm (0.100-in.) Gauge Strip Having a Nominal Base Composition of 1.2% Mn, 0.9% Si, 0.6% Cr, 0.4% Mo

Carbon Content, wt. %	Simulated Coiling Temp., C (F)	Volume % Microconstituent		
		Polygonal Ferrite	Bainitic Ferrite	Martensite-Austenite
0.032	620 (1150)	90.6	3.6	5.8
	595 (1100)	93.4	0.2	6.4
0.065	620 (1150)	87.8	0.5	11.7
0.075	620 (1150)	82.3	5.1	12.6
	595 (1100)	72.4	11.0	16.6
0.091	595 (1100)	57.6	19.1	23.3
	565 (1050)	10.9	65.1	24.0
0.12	595 (1100)	40.2	32.8	27.0
	565 (1050)	19.6	54.9	25.5

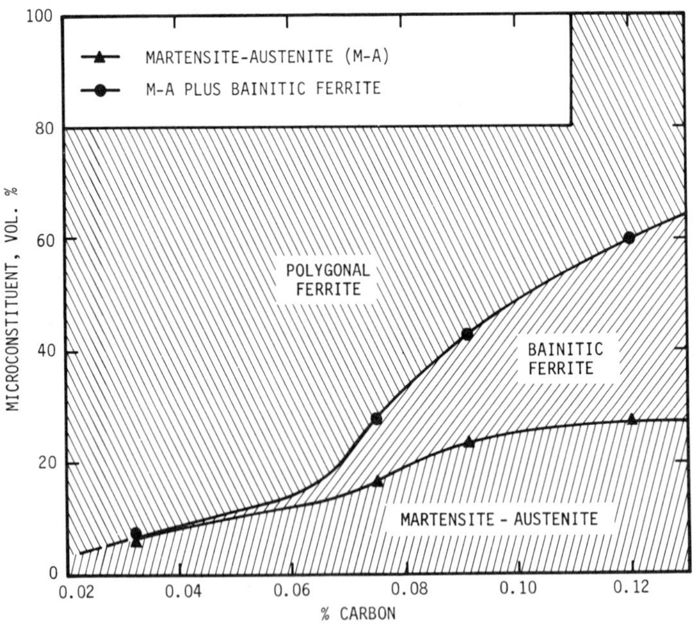

Figure 4 Effect of Carbon Content on Microconstituent Proportions in Laboratory Steels 10-13 Simulated Coiled at 595°C (1100°F)

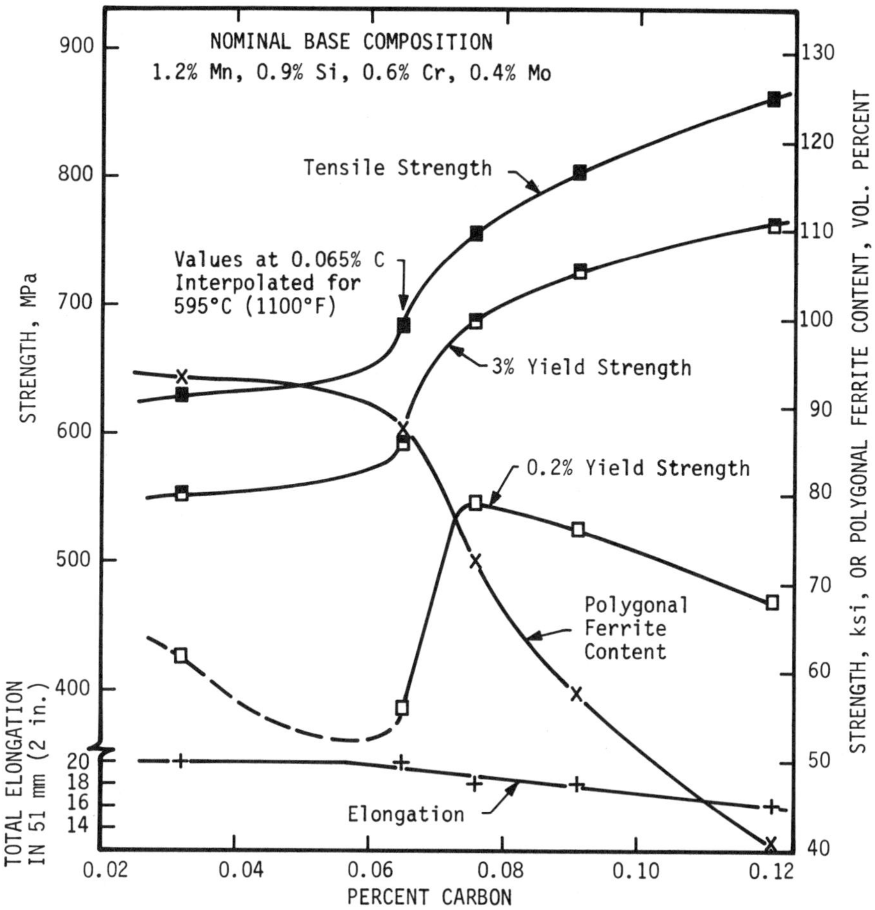

Figure 5 Effect of Carbon on Polygonal Ferrite Content and Tensile Properties of Laboratory Steels 1 and 10-13 Simulated Coiled at 595°C (1100°F)

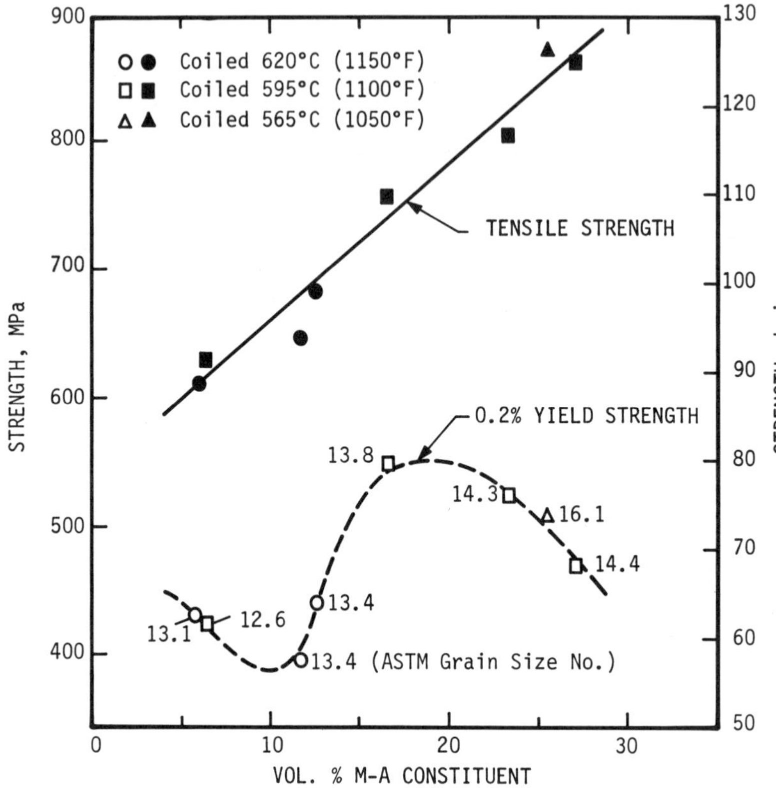

Figure 6 Correlation of Tensile Strength and Yield Strength with Volume Percent Martensite-Austenite Constituent

Effect of Phosphorus in Steels with High Mn/C Ratio. Steels 1 through 6 were employed to ascertain whether variations in phosphorus content would affect the strength-ductility balance of high Mn/C ratio steels. The results are presented in Figure 7 where each point represents the average of duplicate tests. The dashed parallel lines are the VAN-QN band limits reproduced from Figure 2 as a frame of reference. The open points, representing the low-phosphorus steels, formed a group having a superior strength-ductility balance relative to that of the high-phosphorus group. It must be pointed out, however, that the difference between the two groups was somewhat exaggerated by the presence of some cold worked ferrite in several of the high-phosphorus steels. Phosphorus has a strong tendency to raise the Ar_3 temperature, and the finish-rolling temperature employed, 760°C (1400°F), happened to be below the Ar_3 for the high-phosphorus steels, but above the Ar_3 for the low-phosphorus steels. Nevertheless, it was concluded that high phosphorus contents had a tendency to degrade the strength-ductility balance.

Effect of Retained Austenite. An abbreviated study of the effect of retained austenite content was conducted on eleven steels, and the results are presented in Figure 8. Eight of the steels (open points) were Climax Laboratory steels, and three were commercially produced, including two from

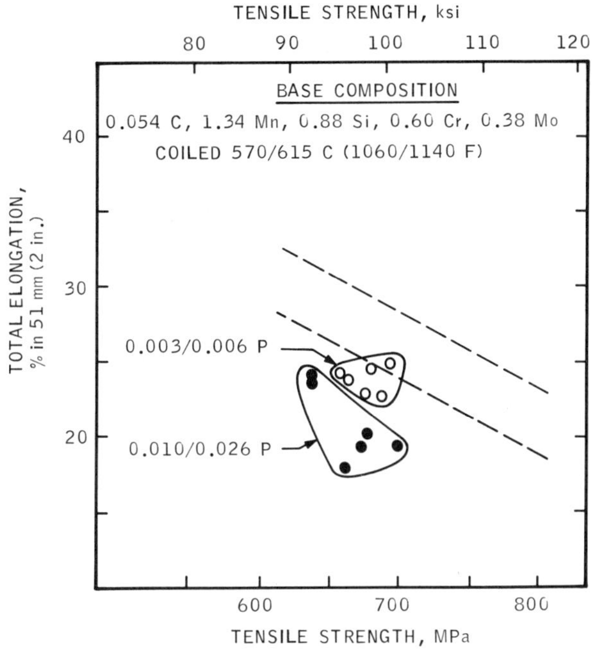

Figure 7 Effect of Phosphorus on Strength-Ductility Balance of Laboratory Steels 3-6

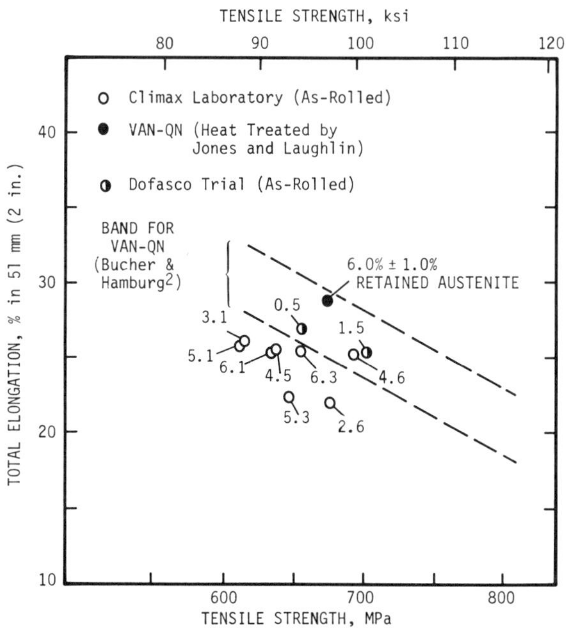

Figure 8 Effect of Retained Austenite on Strength-Ductility Balance

the Dofasco mill trial (results of the Dofasco mill trial will be presented in another section of this paper). The percent austenite values measured ranged from 0.5% (±0.5%) to 6.3% (±1.0%). There was no apparent trend toward improved properties with either high or low retained austenite contents. These data do not permit one to conclude that retained austenite has no effect on properties because no effort was made to vary the austenite content within one steel. On the other hand, the results do lead to the conclusion that high retained austenite contents are not necessary for the attainment of good combinations of strength and ductility. That conclusion is underscored by the two steels from the Dofasco mill trial (half-filled circles in Figure 8) which had the lowest austenite contents observed but exhibited properties that were among the best.

CCT Diagram for 1.2% Mn Steel. The continuous cooling transformation diagram determined for Steel 1 is presented in Figure 9. An austenitizing temperature of 960°C (1760°F) was employed, and no deformation was applied. The absence of deformation resulted in significantly longer transformation start and finish times than would occur in the same steel after being rolled on a hot-strip mill.

The diagram (Figure 9) was characterized by an elongated C-curve for the polygonal ferrite reaction, a delayed pearlite start, and a gap between the ferrite-pearlite region and the bainitic region. The long ferrite nose and delayed pearlite start facilitate the attainment of a large amount of polygonal ferrite on the hot-mill runout table without forming pearlite. The gap provides a coiling temperature range of 85 to 110°C (150 to 200°F) within which no transformation occurs.

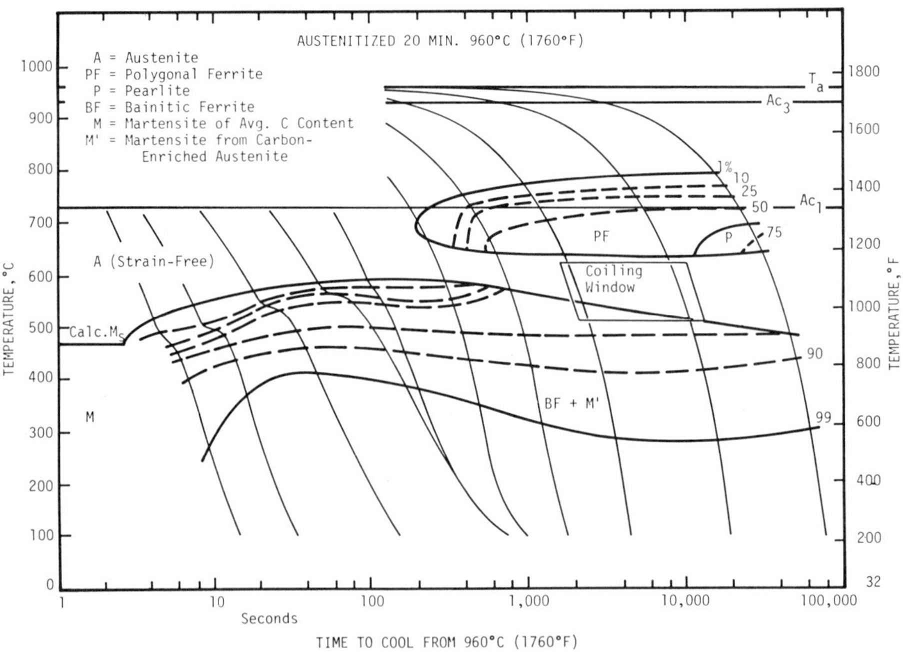

Figure 9 Continuous Cooling Transformation Diagram of Steel 1, Containing 0.065% C, 1.19% Mn, 0.87% Si, 0.38% Mo, 0.61% Cr, 0.064% Al, 0.005% N, 0.01% P, 0.005% S

On the basis of microstructures observed in laboratory-rolled 2.5-mm (0.1-in.) sheets of Steel 1, two conclusions were drawn regarding changes in the CCT diagram that would be found if a diagram were determined for strained austenite. The first was that the ferrite-pearlite region would be shifted to the left by approximately two and one-half orders of magnitude (log cycles). The second was that the contour lines of iso-percent transformation would shift and be more concentrated in the upper portion of the diagram in a manner that, for example, would allow over 85 percent transformation before the beginning of pearlite formation. Both of these effects produced by austenite deformation are very beneficial because they promote the rapid formation of large amounts of polygonal ferrite during the approximately 10 seconds that any given section of a steel strip would spend on the runout table of a commercial mill before entering the coiler.

Discussion of Laboratory Results

The most significant aspect of the laboratory experiments was that they demonstrate an apparent feasibility of producing dual-phase steel in the as-rolled condition on a hot-strip mill. Assuming that the feasibility suggested by the laboratory results proves to be valid under conditions of commercial steel production, and assuming further that a large demand for dual-phase steels develops, the as-rolled approach to dual-phase production will be of great interest to steel mills that do not have heat treating lines available for producing quasi-normalized or intercritically annealed dual-phase steels.

The fact that hot-strip mills vary with respect to (a) the gauge and temperature at entry to the finishing train, (b) rolling speed, and (c) the finish-rolling temperature range means that there may be significant differences from mill to mill in the amount of strain accumulation in the austenite after recrystallization ceases. Consequently, since strain has a strong negative hardenability effect (accelerates ferrite formation), each mill will have to adjust the composition of the steel to be compatible with the specific rolling schedule and runout table cooling rate used. Also, since austenite strain accumulation would affect roll-separating forces, adjustments in the rolling schedule may be necessary to increase the amount of deformation in the recrystallization range and thereby decrease the amount of deformation below the recrystallization range where strain hardening builds up.

The basic question of why dual-phase steels exhibit good ductility at high tensile strength was not resolved in the present investigation. However, two factors that can reduce ductility in as-rolled dual-phase steels were identified, although quantitative measurements of these effects were not made. The two factors were (a) deformed ferrite, produced by finish-rolling below the Ar_3, and (b) bainitic ferrite, produced by excessive hardenability (insufficient polygonal ferrite formation during cooling to the coiling temperature). Both cases involved the introduction of high dislocation density grains into the ferrite component of the ferrite plus martensite/austenite structure. The reduced ductility can therefore be explained on the basis that a significant part of the total capacity of ferrite to "absorb" dislocations before fracture was "used up" before the tensile test was begun. Alternatively, a lower total elongation was obtained because the initial condition of the ferrite was not strain free (dislocation free). It is suggested that maximum ductility can be realized only when the ferrite has a minimum content of (a) dislocations (i.e., containing only those necessarily introduced by the transformation of austenite islands to martensite islands), and (b) unbound atoms of all interstitial and substitutional elements that may degrade the strength-ductility balance.

Dofasco Mill Trial

Following the successful laboratory development program, arrangements were made to produce full size ingots of the most promising composition and process them on a production hot-strip mill. Dofasco has a steel foundry adjacent to its hot-strip mill and can produce heats the size of single ingots [9 tonnes (10 tons)]. This meant that an ideal situation existed for production of a limited number of ingots without the need to use a full heat from the conventional steel making shop, which at Dofasco means 136 tonnes (150 tons) of steel.

Process Details

A twelve-ton Volta electric arc furnace was used to produce two nine-tonne (ten-ton) ingots of dual-phase steel, of the composition felt most likely to give the desired properties. Table IV lists the compositions obtained. The steel was teemed into 610 x 970 x 2300-mm (24 x 38 x 90-in.) ingot molds and mischmetal was added during the teeming for inclusion shape control. In order to avoid potential cracking problems, the ingots were charged while still hot to the soaking pit and then heated slowly to the soak temperature prior to rolling.

Hot rolling was by the following sequence:

1. Slab on a 2 high reversing mill to 280-mm (11-in.) thickness.

2. Surface condition by autoscarfing.

3. Roll on a 2 high reversing mill to 30 mm (1.2 in.) thick.

4. Finish roll to 2.5 mm (0.100 in.) thick on a 4 high 7-stand mill.

5. Cool and coil.

This procedure corresponds to Dofasco's normal practice for wheel rim and spider steels. The hot strip mill rolls directly to coil from ingot and thus no slab reheating was used. Temperatures were nominally within the range recommended as a result of the laboratory experiments.

Prior to rolling the dual-phase steel ingots, a heat of niobium-strengthened 345 MPa (50 ksi) steel was processed to enable a direct comparison of rolling behavior to be made between the two types of steel. After rolling, the coils were pickled and oiled on a continuous process line and finally cut to sheets 860 x 2400 x 2.5 mm (34 x 96 x 0.100 in.). Minimal leveling was used in order to maintain maximum formability. Test samples were taken at frequent intervals (every 10 sheets) throughout the lengths of both coils and mechanical properties, formability and chemical composition checked. Metallography using both optical and scanning electron microscope techniques was done in order to determine the microstructures at each location, and to determine relationships, if any, with mechanical properties.

Table IV Compositions of Steels in Dofasco Trials

	(wt.%)								
	C	Mn	Si	Mo	Cr	Al	N	P	S
Coil 1	0.058	1.62	1.41	0.33	0.67	0.028	0.009	0.010	0.011
Coil 2	0.045	1.20	0.91	0.37	0.61	0.031	0.010	0.013	0.015

Results

As previously mentioned, Table IV gives the composition of the two coils that were produced. Coil 1 contained higher manganese and silicon (1.62% and 1.41%, respectively) than the recommended Climax levels. Coil 2 was as recommended. Mechanical properties were expected to meet the following criteria in order to be called "dual-phase":

(a) Continuous yielding.

(b) 550 MPa (80 ksi) strength after 2% strain.

(c) 655 MPa (95 ksi) tensile strength.

(d) 25% total elongation (min.).

Non dual-phase behavior was the term applied to any material not satisfying one or more of the requirements. Tests on both coils showed that there were substantial areas with dual-phase properties. However, there were also large areas which did not exhibit the required properties. In Coil 1 locations were observed where borderline dual-phase behavior was encountered, i.e., although a distinct yield point elongation did not occur, an inflection at yielding was noted and a little less than 550 MPa (80 ksi) strength was developed after 2% strain. Tables V and VI list properties for each of the types of stress-strain curve described.

Table V Typical Properties from Coil 1

	Dual-Phase	Transition	Non Dual-Phase
0.2% Yield Strength, MPa (ksi)	450 (65)	485 (70)	495 (72)
Yield Point Elongation, %	--	--	1
Strength after 2% Strain, MPa (ksi)	545 (79)	--	--
Tensile Strength, MPa (ksi)	690 (100)	655 (95)	640 (93)
Total Elongation in 51 mm (2 in.), %	27	28	27

Table VI Typical Properties from Coil 2

	Dual-Phase	Non Dual-Phase
0.2% Yield Strength, MPa (ksi)	450 (65)	415-485 (60-70)
Yield Point Elongation, %	--	1-2
Strength after 2% Strain, MPa (ksi)	545 (79)	--
Tensile Strength, MPa (ksi)	655 (95)	535-585 (78-85)
Total Elongation in 51 mm (2 in.), %	25	25-35

Metallographic examination of the steel showed that, as expected, dual-phase behavior was associated with a mixed structure of polygonal ferrite and martensite islands. Surprisingly, though, there was also a substantial amount of grain boundary carbide and, in some cases, bainitic ferrite was also present. Large variations in the relative amounts of each phase did not give associated changes in the mechanical properties. Polygonal ferrite contents ranged from 70% to 90%, bainitic ferrite from 0% to 25%, and up to 10% martensite was present.

Regions of the coils with non dual-phase properties contained similar quantities of polygonal ferrite, less bainitic ferrite, much more grain boundary carbides, small amounts of pearlite and, in some cases, small quantities of martensite. Figures 10, 11 and 12 give typical examples of the structures associated with dual-phase behavior and with non dual-phase properties.

During the rolling of the steels it was noted that they were harder than the 345 MPa (50 ksi) yield strength niobium steels rolled just prior to these. The mill loads were considered similar to those encountered with 415 MPa (60 ksi) steels.

Discussion of Mill Trial Results

The results obtained from the mill trial show that the concept developed in the laboratory has promise, but that more development work is necessary because of the erratic performance. The variability was probably the result of different cooling conditions (either cooling rate or coiling temperature or a combination of the two). Figures 13 and 14 show hypothetical cooling curves which could give rise to the structures observed. Where non dual-phase properties were the result, cooling rates could have been too slow and thus, resulted in pearlite formation. Alternatively, coiling temperatures may have been too high, giving the same effect. Formation of pearlite reduces the

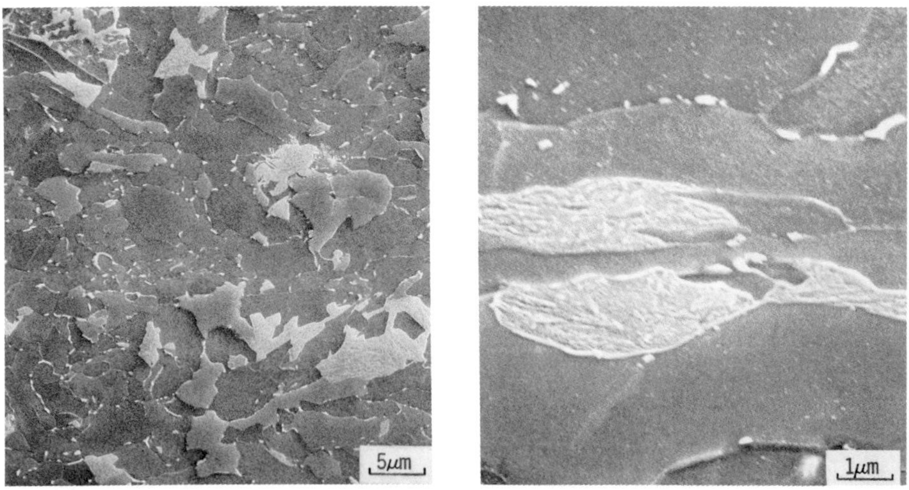

Figure 10 SEM Micrographs Showing Structure that Produced Dual-Phase Behavior in Coil 1

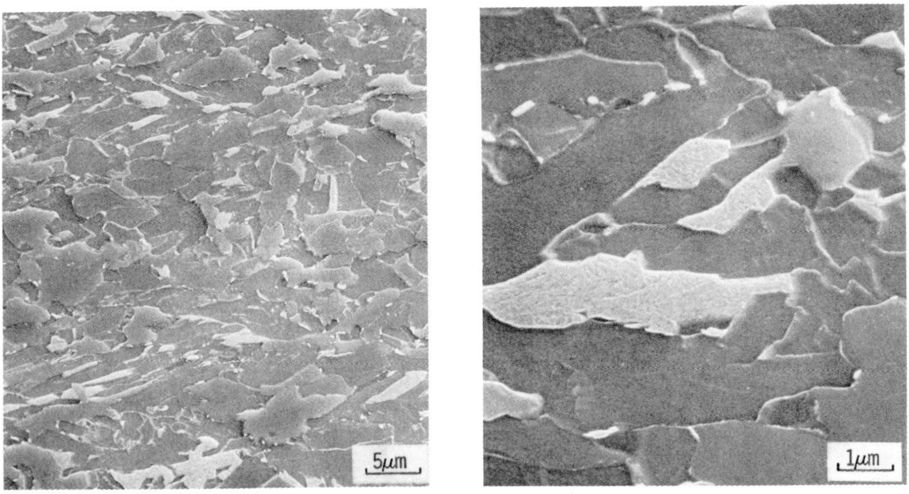

Figure 11 SEM Micrographs Showing Structure that Produced Dual-Phase Behavior in Coil 2

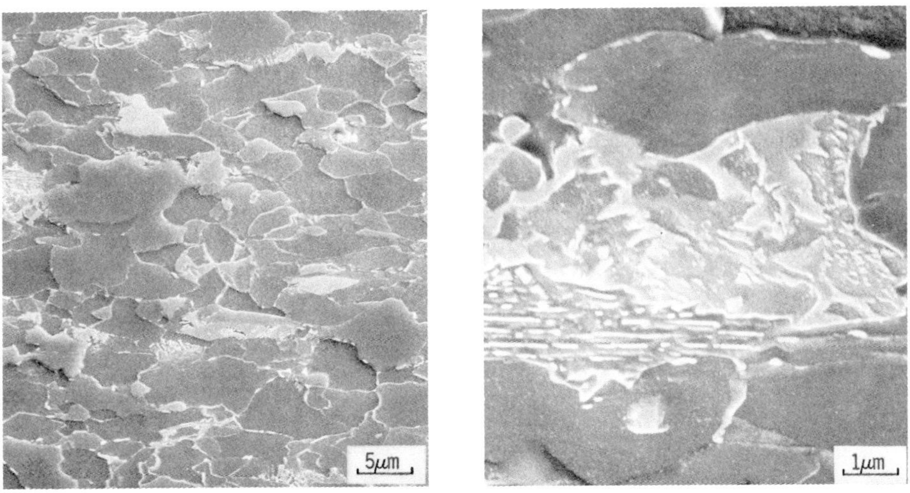

Figure 12 SEM Micrographs Showing Structure that Resulted in Non Dual-Phase Behavior in Coil 2

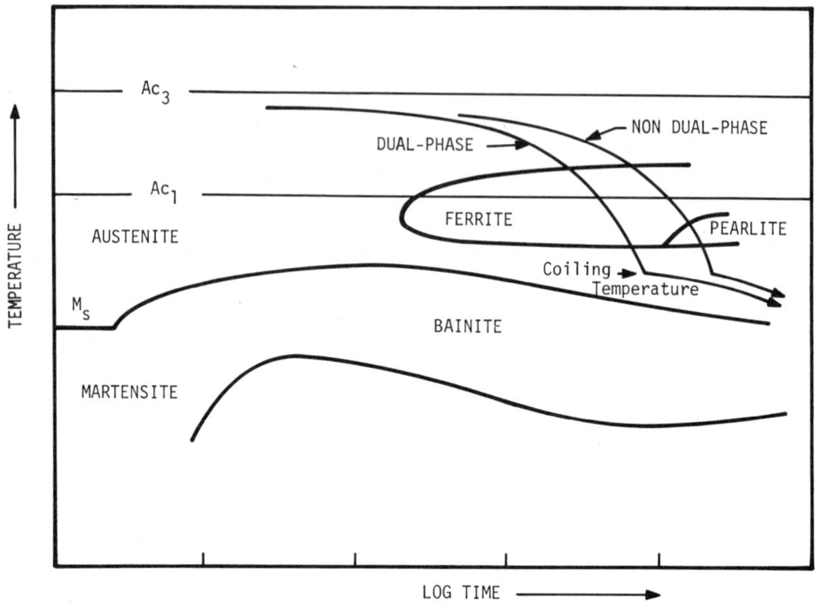

Figure 13 Schematic Representation Illustrating How a Non Dual-Phase Microstructure Could Have Resulted from an Excessively Slow Cooling Rate on Runout Table

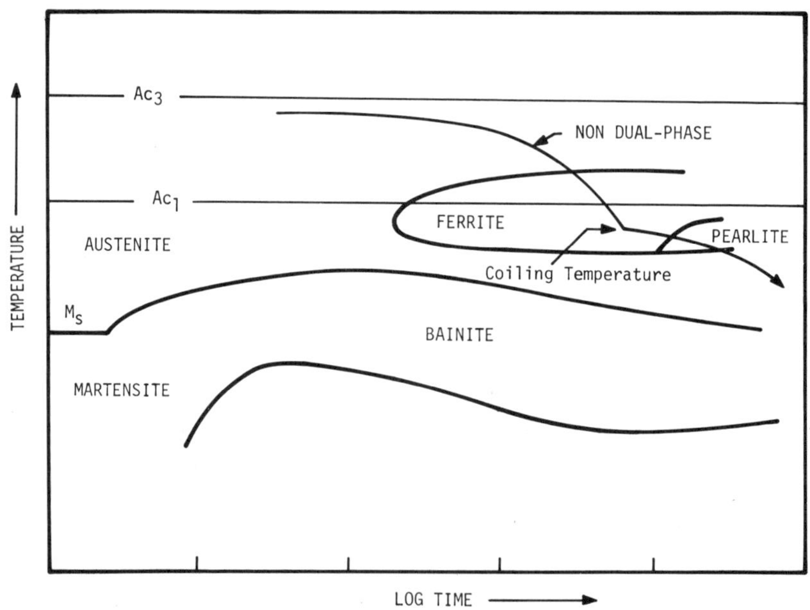

Figure 14 Schematic Representation Illustrating How a Non Dual-Phase Microstructure Could Have Resulted from an Excessively High Coiling Temperature

amount of remaining austenite and, consequently, the amount of martensite in the final structure. Dual-phase properties were obtained when pearlite was avoided. The proposed cooling conditions are, however, hypothetical and have not been confirmed by measurement. Subsequent mill trials will check the theories.

One other point evident from the trial is the insensitivity of the properties to marked changes in microstructure such as amount of martensite and bainitic ferrite. This suggests that, once the correct cooling conditions have been established, the steel is likely to be insensitive to normal variations in production conditions.

Future work will be aimed at eliminating property variability by changes in composition and process conditions. In subsequent trials, BOF heats may be made. Since these contain less phosphorus and nitrogen there is a possibility that the ductility of the steel will be improved. More development is planned, therefore, since the prospects of success are high. Preliminary results from a subsequent trial look very promising since dual-phase behavior was observed consistently throughout the lengths of two 9-tonne (10-ton) coils.

Summary and Conclusions

The combined results of a laboratory development program and a commercial mill trial have established the possibility that a dual-phase steel may soon be in commercial production as an as-rolled product from hot-strip mills. The laboratory program, conducted at the Climax Molybdenum Laboratory, led to the development of a steel with a CCT diagram exhibiting special features, including (a) an elongated C-curve for the polygonal ferrite reaction with a delayed pearlite-start time and (2) a large gap separating the ferrite/pearlite region from the bainite region. The first feature permits large amounts of polygonal ferrite to be obtained without forming pearlite during cooling on the runout table. The second feature provides a large coiling temperature range within which no transformation occurs.

The commercial mill trial was conducted by Dominion Foundries and Steel, Ltd., (Dofasco) and involved two 9-tonne (10-ton) electric furnace heats. The heats, cast as single ingots and rolled on the Dofasco hot mill to single coils, both exhibited dual-phase properties and structures in some regions of the coils. Microstructural variations observed along the length of each strip were correlated with the properties, and it appeared that pearlite formation was the cause of non dual-phase properties. It was concluded that changes in the runout cooling conditions would probably lead to uniform, dual-phase steel coils in future trials.

The following conclusions were drawn from the Climax Laboratory data and the Dofasco mill trial results:

1. The possibility of producing dual-phase steel as an as-rolled product directly off the hot-strip mill has been demonstrated in laboratory experiments and in a commercial mill trial.

2. Based on the combined results of the laboratory studies and the mill trial, the recommended steel composition and hot-mill parameters for producing as-rolled dual-phase steel are as follows:

Steel Composition (Wt. %)

	C	Mn	Si	Cr	Mo	Al
Range:	0.065 max.	1.0/ 1.2	0.9/ 1.1	0.4/ 0.6	0.35/ 0.45	0.02 min.
Preferred:	0.05	1.1	1.0	0.5	0.4	0.05

N, P and S as Low as Possible; Inclusion Shape Control with Rare Earth (or Possibly Zirconium) Additions

Processing Parameters

Slab Soaking Temperature - - -	1260/1315°C (2300/2400°F)
Finish-Rolling Temperature - -	845/900°C (1550/1650°F)
Coiling Temperature - - - - -	540/595°C (1000/1100°F)

3. Retained austenite contents that were determined by X-ray diffraction on eleven dual-phase steels, including three commercially produced steels, exhibited no correlation with the strength-ductility (tensile strength vs % elongation) balance.

References

1. Rashid, M. S., "GM 980X - Potential Applications and Review," SAE Paper 770211, Presented at meeting of Society of Automotive Engineers in Detroit, Michigan, February 28, 1977.

2. Bucher, J. H. and Hamburg, E. G., "High Strength Formable Sheet Steel," SAE Paper 770164, Presented at International Automotive Engineering Congress and Exposition, Detroit, Michigan, March 1977.

3. Hamburg, E. G., Cryderman, R. L. and Butler, J. F., "Method of Producing a High Strength Steel Having Uniform Elongation," U.S. Patent 4,033,789, July 5, 1977.

4. Cias, W. W., "Phase Transformation Kinetics and Hardenability of Medium-Carbon Alloy Steels," Climax Molybdenum Company, Greenwich, Connecticut.

5. Cias, W. W., "Phase Transformational Kinetics of Low-Carbon, Boron-Treated Steels," Climax Laboratory Report, February 15, 1971, Unpublished Research.

DISCUSSION

S. K. Banerji, Foote Mineral Company, Route 100, Exton, Pennsylvania 19341

It appears that hardenability is one of the key factors in achieving a "dual-phase" structure, particularly in the as-rolled product. It also seems that large amounts of alloying elements such as 0.6 Cr, 0.4 Mo, etc., are used only to get a certain desired hardenability. What about using boron for this purpose? The cost benefits to be derived in that case is self evident.

Authors' Reply

Two kinds of hardenability are needed — one for avoiding pearlite during cooling to the coiling temperature and the other for avoiding transformation of the remaining austenite islands to bainite during slow cooling in the coil. Boron has a strong inhibiting effect on ferrite nucleation and thereby, indirectly, it delays pearlite formation. However, the slow-down in ferrite formation works against attainment of the desired 85% ferrite during rapid cooling on the runout table. Boron's hardenability effect during slow cooling in the coil is not known.

The success attained with the Mn-Si-Cr-Mo steel is attributed to the unusual effects of the Si-Cr-Mo trio in (1) allowing rapid ferrite formation while preventing pearlite nucleation on the runout table, (2) creating the gap between the upper and lower portions of the CCT diagram, and (3) inhibiting bainite formation during slow cooling in the coil.

Eight boron-containing steels were included in the Climax Laboratory exploratory survey of steels for the as-rolled dual-phase program. They were found to be unsatisfactory because of insufficient ferrite formation during cooling from the finish-rolling temperature to the coiling temperature.

DISCUSSION

A. R. Marder, Research Department, Bethlehem Steel Corporation, Bethlehem, Pennsylvania 18015

In determining the effect of boron in "dual-phase" alloys the mechanism of boron hardenability must be considered. In general, theories on the boron effect are based on the solution of boron in austenite and its segregation to austenite grain boundaries, where it retards the grain boundary nucleation reaction, i.e., ferrite and pearlite (see e.g., A. Brownrigg, "Boron in Steel — A Literature Review 1956-1972," J. Australian Inst. of Metals, Vol. 18, No. 3, 1973, p. 124; S. R. Keown, "Boron in Steel", Scandinavian Journal of Metallurgy, Vol. 2, 1973, p. 59). Several factors indicate that boron will not be effective in intercritical annealing. First, it is quite doubtful that the boron precipitates formed in the austenite to increase hardenability (e.g., $V_4(BC)_3$) would be soluble at the intercritical annealing temperature of approximately 1450°F. Second, during intercritical annealing the most important factor is segregation of solute elements into the austenite where hardenability is further improved. In both cases, boron would not be effective in producing the "dual-phase" structure and properties during intercritical annealing. Furthermore, at our laboratory we have intercritically annealed boron-containing vanadium steels and found no beneficial effect. In considering as-rolled structures, boron will have a tendency to move the ferrite and pearlite nose to the right in the CCT diagram thus allowing for the formation of diffusionless structures such as martensite, bainite and acicular ferrite. We have experienced such structures in as-rolled boron steels and consequently the typical "dual-phase" properties were not obtained.

DEVELOPMENT OF QUENCHED AND TEMPERED 760 MPa GRADE STEEL SHEETS WITH SUPERIOR BENDABILITY

Takashi Matsuoka, Morio Takahashi*,

Norihiko Nozaki**, Yasutaka Okada

Sumitomo Metal Industries, Ltd., Central Research Laboratories,

Amagasaki City, Japan

Abstract

Various factors which affect bendability, were investigated to develop quenched and tempered steel sheets with a yield strength level of 760 MPa and superior bendability. While minor changes of alloying elements such as carbon, manganese, chromium, and so forth affect bendability to a small extent, additional elements such as sulfur, aluminum, and calcium have a large effect on bendability. Calcium is especially effective in improving bendability. The effects of the volume and shape of non-metallic inclusions are discussed.

* Head Office, Sumitomo Metal Industries, Higashiku, Osaka City, Japan

** Kashima Works, Sumitomo Metal Industries, Kashima, Ibaragi Prefecture, Japan

Introduction

High-strength steel sheets with yield strength levels lower than 550 MPa have been commercially produced for automobiles. One of the most essential properties of these sheets is a ductility good enough to endure cold-forming. To date, steel sheets which have a yield strength level higher than 550 MPa and are applicable to press-forming have not yet been available. The main reasons are: firstly, such steel sheets are difficult to produce due to the high deformation-resistance during hot rolling, and secondly, their press-forming is also critical due to low ductility and large spring-back. A large press-capacity is necessary to stamp these sheets.

In order to avoid these difficulties, sheets with low strength are conventionally press-formed and subsequently heat-treated to a yield strength level of 760 MPa. This process, however, has some disadvantages, such as surface decarburization and distortion of pressed parts, which are caused by heat-treatment (1).

The object of the present investigation was to develop a steel sheet with a minimum yield strength of 760 MPa to be applicable to the frame of a heavy-duty truck. Preliminary experiments disclosed that the bendability of quenched and tempered steel sheets is superior than that of hot rolled bainitic steel sheets at a yield strength level of 760 MPa. Based on this fact, examinations on quenched and tempered steel sheets have been carried out.

Experimental Procedure

Steels A to K shown in Table I were melted in induction furnaces and cast into 100 kg or 2-ton ingots. Ingots were forged and hot rolled from 30 mm to 6.0 mm in thickness. Cross rolling (LT rolling), using steel A, means hot rolling which is performed by making a 90-degree-turn after each pass, in contrast to an ordinary rolling (LL rolling); that is, in one direction.

All the steel sheets were heat-treated between 900°C and 930°C for 30 minutes, quenched into water and tempered between 600°C and 650°C for 30 minutes to get an approximately constant yield strength of 760 MPa.

Tensile, bend and press bend specimens were sectioned as shown in Table II and Figure 1. Bend specimens were bent to 180 degrees, using an apparatus shown in Figure 2 and bendability was evaluated with a "minimum bend radius" criterion, indicating the minimum radius obtained without any crack formation among the various radii of bend tests. Press bend was performed on a hydraulic press, using an apparatus shown in Figure 3.

Non-metallic inclusions were evaluated in accordance with JAPANESE INDUSTRIAL STANDARD (G 0555) and the volume fractions of three kinds of inclusions shown in Table III were measured.

Experimental Results

Effects of Alloying Elements

Effects of alloying elements on the bendability were examined by using steels A, B, C and D with various amounts of carbon, manganese and chromium and so forth. The results are shown in Table IV and Table V. As shown in

Table 1 Chemical compositions of steels (wt %)

Steel	C	Si	Mn	P	S	Cr	Mo	V	Sol.Al	Reference
A	0.14	0.41	1.50	0.014	0.008	0.48	0.12	0.04	0.018	Induction melt (2 ton)
B	0.11	0.30	1.01	0.016	0.009	2.18	0.50	-	0.022	Effect of chemical compositions.
C	0.08	0.28	1.50	0.014	0.009	0.48	-	-	0.013	
D*	0.13	0.32	1.47	0.014	0.010	1.07	0.28	-	0.020	* Cu: 0.25%, Ni: 0.26%
E	0.15	0.34	1.43	0.006	0.007	0.49	0.11	0.03	0.030	Effect of heating temperature.
F	0.20	0.38	1.44	0.002	0.005	0.51	0.12	-	0.025	Induction melt (100 kg)
G	0.17	0.38	1.51	0.009	0.023	0.52	0.11	-	0.027	Effect of sulfur content.
H	0.18	0.38	1.45	0.008	0.040	0.51	0.11	-	0.027	
I	0.19	0.35	1.44	0.002	0.005	0.50	0.10	-	0.001	Induction melt (100 kg)
J	0.19	0.40	1.44	0.002	0.005	0.50	0.10	-	0.062	Effect of Sol.Al content.
K	0.20	0.35	1.45	0.002	0.005	0.50	0.10	-	0.088	

Table II. Sampling Direction of the Specimens

Specimen	Sampling Direction
Tensile	L
Bend	T
Press	L and T

L : Longitudinal to the rolling direction.

T : Transverse to the rolling direction.

Table III Type of non-metallic inclusions

Type	Shape of non-metallic inclusion	Composition
A	Elongated to the rolling direction	manganese sulfide, silicate
B	Scattered in the rolling direction	aluminium oxide
C	Irregularly dispersed	granular oxide

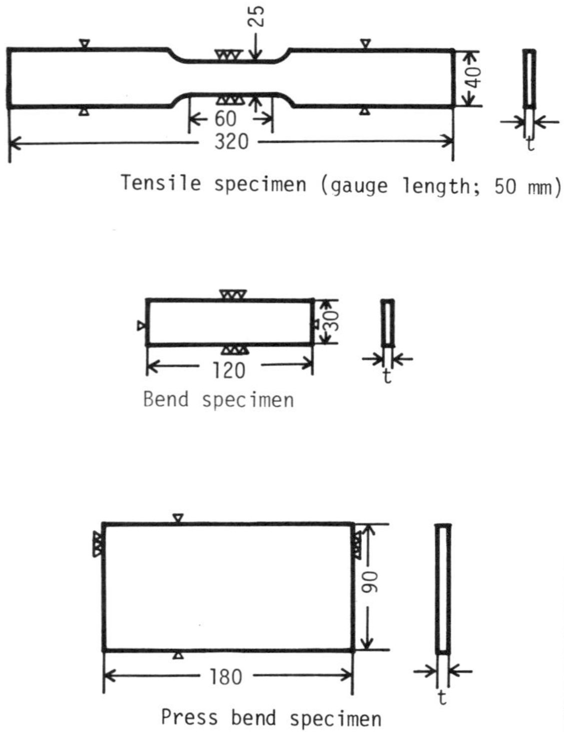

Fig. 1 Specimens (mm)

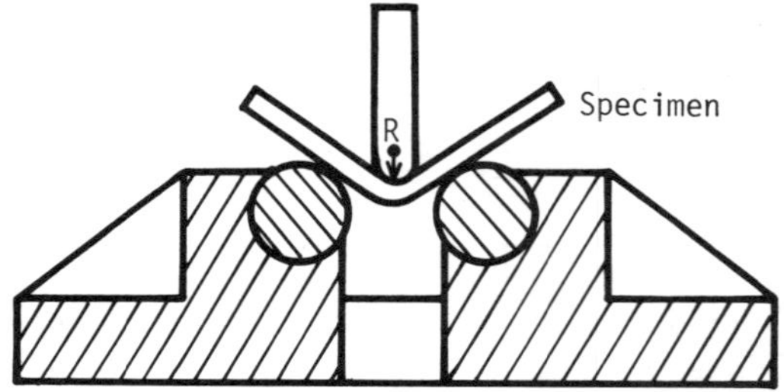

Fig. 2 Apparatus for bend test

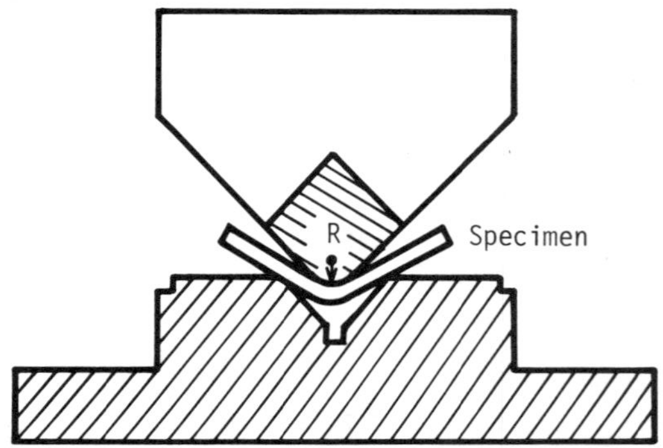

Fig. 3 Apparatus for press bend test

Table IV, their yield strengths are in the level of about 760 MPa and other tensile properties are also similar. Only A- and B-type inclusions were observed in these steels. It is characteristic that their bendability is also similar regardless of the different amounts of alloying elements. This fact demonstrates that minor changes in these alloying elements do not seriously affect bendability. Accordingly the base composition of the steel was selected to be 0.13% to 0.2% C, 0.3% Si, 1.4% Mn, 0.5% Cr, 0.1% Mo*.

Effects of Hot Rolling Method

Generally, bendability in the rolling direction (L direction) is superior to the transverse direction (T direction). Comparison of the bendability of LT rolling with LL rolling (specimens were taken in the (final) direction of rolling) is examined as shown in Table VI. Steel sheets of LT rolling show superior bendability to some extent, but the level of improvement is not remarkable.

Effects of Heating Temperature

Figure 4 shows the effects of heating temperature of 900°C, 1000°C and 1200°C prior to quenching on tensile properties and bendability of steel E.

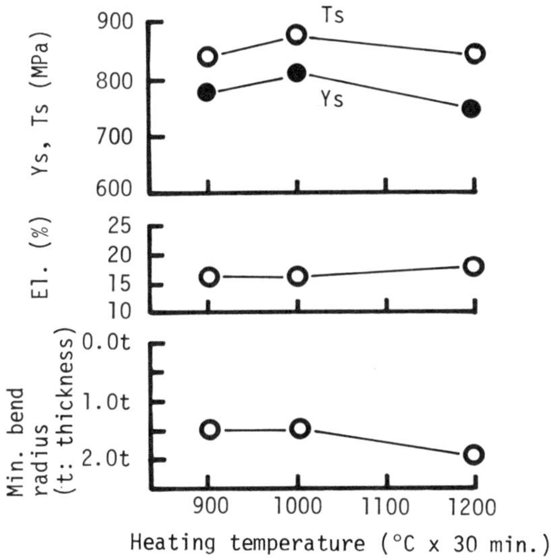

Fig. 4 Effects of heating temperature on the mechanical properties. (tempered at 600°C for 30 min.)

* The composition was determined in reference to the requirement of toughness that Charpy 50% brittle fracture transition temperature, vTs, should be lower than -40°C.

Table IV Mechanical properties of various steels

Steel		Ys MPa	Ts MPa	El. %	Min. bend radius t: thickness	Press bend t: thickness
A	Cr-Mo-V	756	784	23	1.0t	1.5t good
B	High Cr-Mo	762	821	23	0.5t	1.5t good
C	Low C-Cr-Mo	756	804	24	1.0t	1.5t good
D	Cu-Ni-Cr-Mo	746	785	24	1.0t	1.5t good

t: 6.0 mm

Table V Non-metallic inclusions (volume fraction, %)

Steel	Type			Total
	A	B	C	
A	0.042	0.033	0.000	0.075
B	0.021	0.025	0.000	0.046
C	0.033	0.017	0.000	0.050
D	0.017	0.013	0.000	0.030

Table VI Effects of cross rolling on the bendability

Steel	Hot rolling	Ys MPa	Ts MPa	El. %	Min. bend radius (t: thickness)
A	LL rolling	775	826	19	1.5t
A	LT rolling(L/T = 0.8**)	755	803	20	1.0t

* After hot rolling, sheets, 6 mm thick, were quenched and tempered.

** Ratio of hot rolling reduction. (Longitudinal/Transverse direction.)

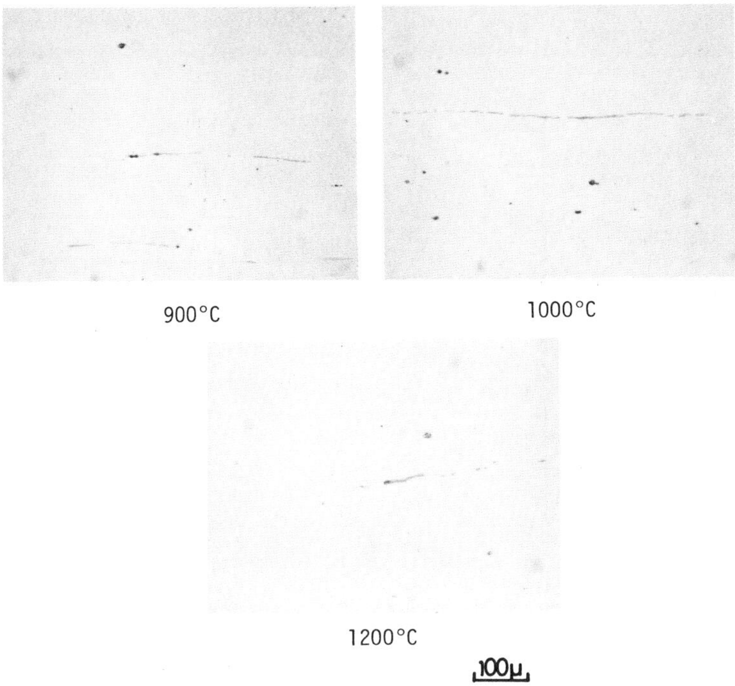

Fig. 5 Effects of heating temperature on the shape of non-metallic inclusions.(Steel E)

Figure 5 shows the change in the shape of the non-metallic inclusions associated with the variation of the heating temperature. Long elongated inclusions are observed in the steel sheets heated at 900°C. The inclusions dissociate into small particles at 1000°C. At the highest temperature of 1200°C, these inclusions start to cohere and their tails become round, but their length is still larger in comparison to their width. Bendability is not improved in spite of the changes in the inclusions at the elevated temperatures. This is because the inclusions heated at 1000°C and 1200°C are still aligned in a same line along the rolling direction even though they dissociate into small particles.

Effects of Sulfur and Aluminum

Among the various additional elements, sulfur and aluminum are the main elements to compose A-type and B-type inclusions, respectively. Figure 6 indicates the effect of sulfur content on the bendability, using the steels F, G and H. Bendability is remarkably improved by the decrease of sulfur content, especially less than 0.010%.

The correlation of acid soluble aluminum (Sol. Al) content to the bendability is shown in Figure 7, using the low-sulfur steels I, J and K. Addition of aluminum improves bendability, but bendability deteriorates, when Sol. Al is more than 0.08%. In a range of 0.025 to 0.060% Sol. Al, a minimum bend radius of 0.5 t (t; thickness of the steel sheet) is obtained.

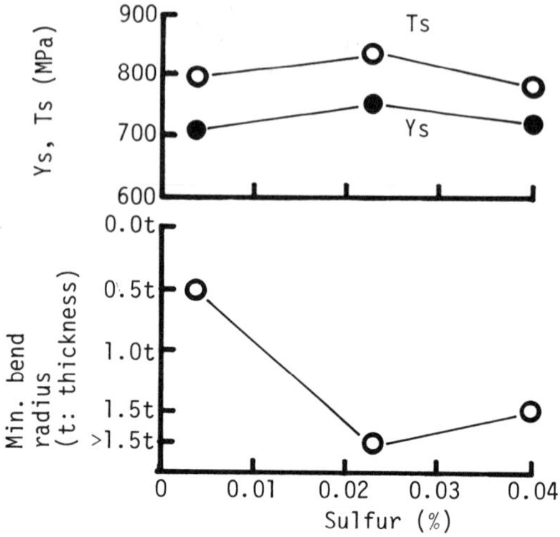

Fig. 6 Effects of sulfur content
(0.19%C-0.35%Si-1.4%Mn-0.5%Cr-0.1%Mo)

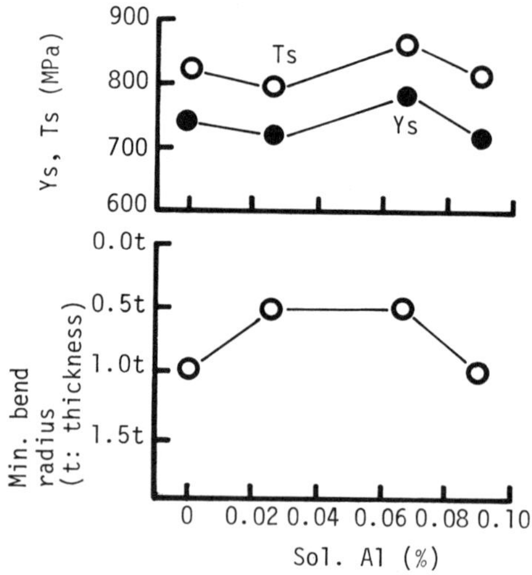

Fig. 7 Effects of Sol. Al content
(0.19%C-0.35%Si-1.45%Mn-0.5%Cr-0.1%Mo)

Steel	L	M	N
Sol.Al (wt %)	0.024	0.015	0.042
Ca (wt %)	0.0005	0.0042	0.0022
Ys MPa	803	770	797
Min. bend radius. (t: thickness)	1.0t	1.0t	0.0t
Shape of non-metallic inclusions 200 µm	 Steel M		 Steel N

Fig. 8 Mechanical properties and shape of non-metallic inclusions of calcium added steel sheets.

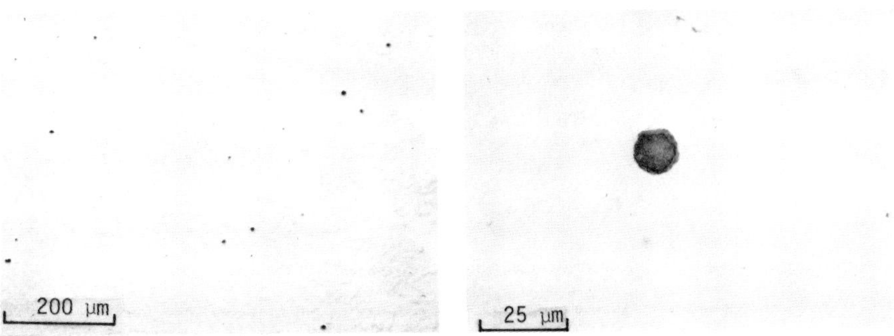

Fig. 9 Non-metallic inclusions of calcium treated steel sheet. (steel N, 6 mm thick.)

Effects of Calcium Additions

The chemical compositions of the calcium-treated steels, L, M and N are shown in Table VII. In the case of steel L and M, containing relatively small amounts of Sol. Al or calcium, bendability is not remarkably improved, as shown in Figure 8. The inclusions are not always spheroidized completely, though they are not so much elongated as those shown in Figure 5.

Steel N, manufactured in a 25-ton electric furnace in the steelmaking shop, has enough Sol. Al (0.046% Al) and calcium (0.0022% Ca). The sulfur content (0.008% S) is not extremely low, but most of the non-metallic inclusions consisted of spherical ones of C-type. A-type inclusions were few in number and B-type ones were not observed, as shown in Table VIII and Figure 9.

Mechanical properties are shown in Table VIII. Ten press bend specimens were classified into three classes, A, B and C, depending on the degree of cracking. Bendability and press bendability are remarkably good as shown in Figure 10.

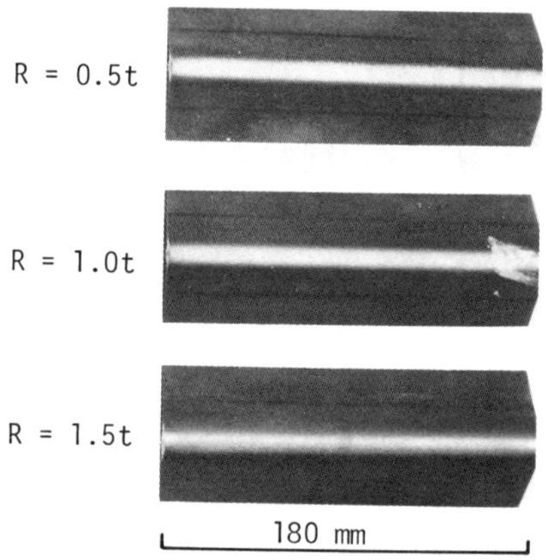

Fig. 10 Result of press bend test.
(R: bend radius, t: thickness of the sheets)

Fig. 11 Frame (steel N; 6.0 mm thick)

Table VII Chemical compositions of calcium added steels (wt %)

Steel	C	Si	Mn	P	S	Cr	Mo	V	Sol.Al	Ca	Reference
L	0.16	0.33	1.45	0.006	0.007	0.49	0.11	0.03	0.024	0.0005	Induction melt (100 kg)
M	0.18	0.35	1.42	0.010	0.010	0.50	0.11	-	0.015	0.0042	Induction melt (100 kg)
N	0.13	0.37	1.32	0.011	0.008	0.61	0.10	0.03	0.046	0.0022	25 ton electric furnace melt

Table VIII Non-metallic inclusions and mechanical properties of 760 MPa grade steel sheets (steel N)

Non-metallic inclusion* %		Thickness mm	Sampling direction	Ys MPa	Ts MPa	El. %	Bend test Min. bend radius (t:thickness)	Press bend test** Bend radius 0.5t			1.0t		
								A	B	C	A	B	C
A	0.008	6.0	L	797	852	22	-	10	0	0	10	0	0
B	0.000		T	796	850	19	0.0t	9	1	0	10	0	0
C	0.071	8.0	L	760	815	22	-	6	4	0	10	0	0
Total	0.079		T	761	818	20	0.5t	6	4	0	10	0	0

* volume fraction

** Degree of crack

A	good (no crack)
B	micro-crack
C	crack

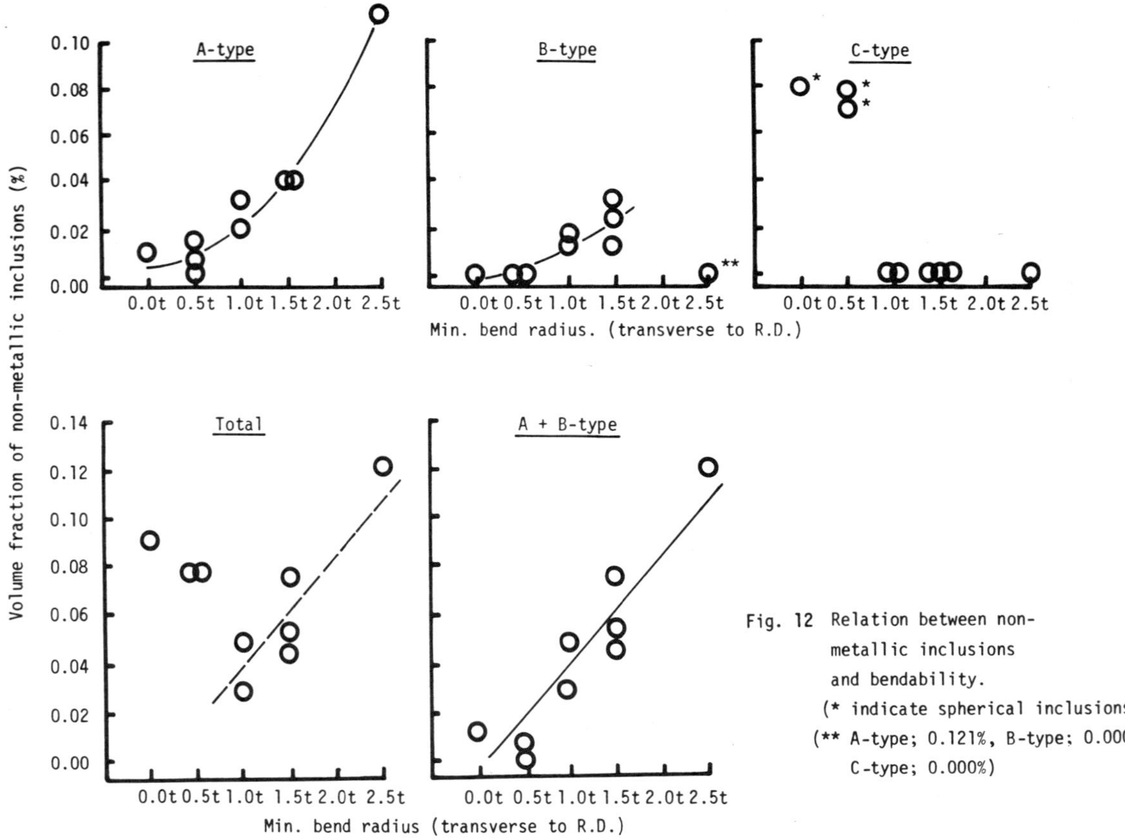

Fig. 12 Relation between non-metallic inclusions and bendability.
(* indicate spherical inclusions)
(** A-type; 0.121%, B-type; 0.000% C-type; 0.000%)

Sulfur (%) Sol. Al (%) Calcium (%)		0.008 <0.01 -	0.008 0.03 -	0.008 0.046 0.0022
Shape of non-metallic inclusions		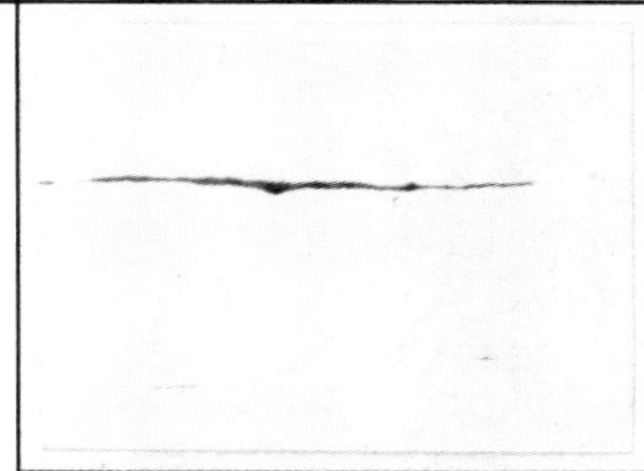		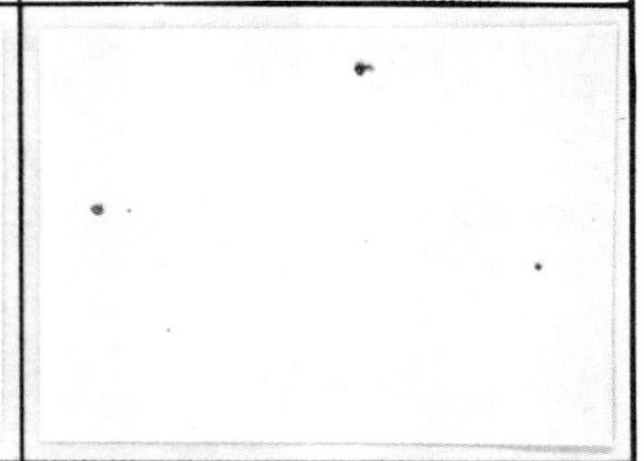
Volume fraction	A B C	0.121 0.000 0.000	0.042 0.033 0.000	0.012 0.000 0.079
	Total	0.121	0.075	0.091
Bend test		2.5t good	1.5t good	0.0t good
Press bend test		>2.0t	1.5t good	0.5t good

Fig. 13 Relation between non-metallic inclusions and bendability

These sheets, 6 mm in thickness, were successfully stamped to a frame for a heavy-duty truck at Press Kogyo Company using the die for production, as shown in Figure 11. In a commercial run, this frame is made of 400 MPa-grade steel and is formed to a bend radius of 1.5 times the thickness.

Discussion

While alloying elements such as carbon, manganese, chromium and so forth, affect bendability to a small extent, aluminum, sulfur and calcium have a much larger effect. These latter elements have a major influence on the volume and shape of non-metallic inclusions, which, in turn, are closely related to bendability.

Figure 12 shows the relation between the minimum bend radius and the volume fraction of non-metallic inclusions of each type. Bendability deteriorates with an increase in A- and B-type inclusions, but it is independent of C-type inclusions as far as the experiment concerns. Therefore, the volume fraction of A- plus B-type inclusions is closely correlated to the bendability.

According to the micrographs shown in Figure 13, the addition of aluminum is effective in decreasing A-type inclusions, because it diminishes the silicate inclusions of A-type. Inclusions of aluminum oxide of the B-type are likely to cause surface defects and surface cracks in bending, as shown in Figure 14. Accordingly, excessive additions of aluminum should be avoided to prevent surface defects.

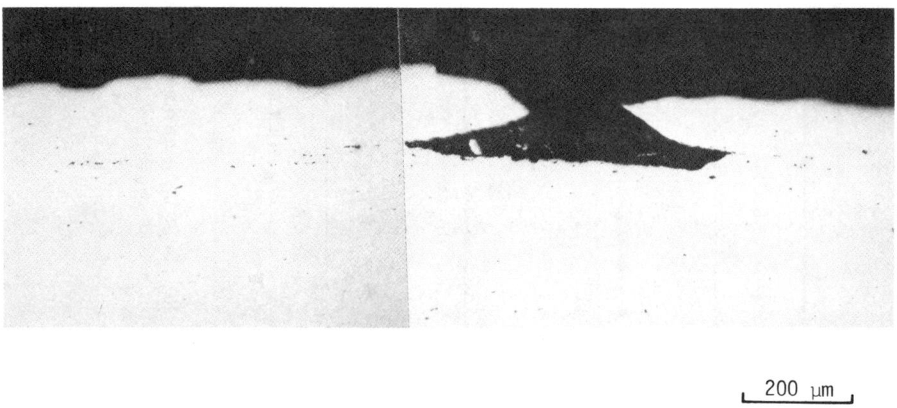

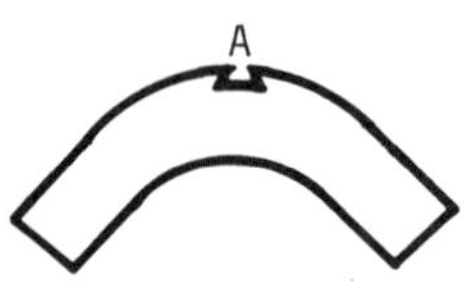

Fig. 14 The cross section view of cracked portion A of a bend specimen.

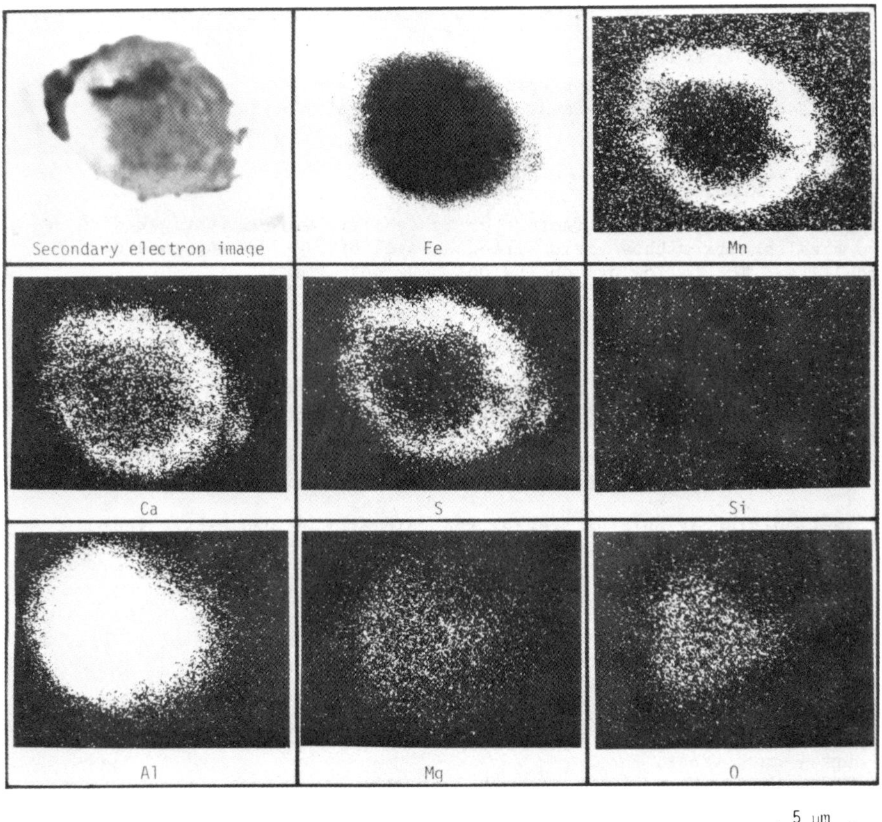

Fig. 15 Electron probe micro-analysis of a spherical inclusion.

The addition of calcium is markedly effective in decreasing A- and B-type inclusions and making the inclusions spherical as shown in Figure 13. Figure 15 is a result of an electron microprobe analysis on one of these spherical inclusions. It is clearly observed that manganese, calcium, aluminum and sulfur are component elements of the spherical inclusion. Shiraiwa and Fujino (2) observed similar non-metallic inclusions in a calcium-treated free-cutting steel by adding aluminum and explain this effect on the basis of the following equation calculated for a temperature of 1000°C using data from Kubaschewski (3):

$$3MnS + 3CaO + 2Al \rightleftarrows 3CaS + Al_2O_3 + 3Mn - 88.7 \times 10^3 \text{ (Kcal)} \qquad [1]$$

These results suggest that appropriate aluminum addition is important in the optimization of the effectiveness of calcium additions.

Conclusions

The effects of various factors on bendability were investigated to develop steel sheets with a yield strength level of 760 MPa and good cold-formability. The following conclusions have been made:

(1) Alloying elements and hot rolling method affect bendability to a small extent.

(2) Increasing the heating temperature prior to quenching modifies the shape of the elongated inclusions, but bendability is not improved.

(3) The amount of sulfur and aluminum is a major factor controlling the volume of A- and B-type inclusions and has marked effects on bendability.

(4) The addition of calcium reduces the volume of A- and B-type inclusions and it makes the inclusions spherical. Bendability is markedly improved, and 180 degree bends bent back on themselves are possible in steels with high yield strength levels of 760 MPa.

(5) Bendability is closely correlated to the volume fraction of A- plus B-type inclusions but is independent of C-type inclusions.

(6) These developments have been utilized to develop high-strength steel sheets which have been successfully press formed into a frame for a heavy-duty truck.

References

(1) M. W. Johnson: Microalloying 1975 Proceedings of an International Symposium on High-Strength Low-Alloy Steels, Published by Union Carbide Corporation, 1977, p. 643.

(2) T. Shiraiwa and N. Fujino: Tetsu to Hagane 57 (1971) 1990.

(3) O. Kubaschewski and E. L. Evans: Metallurgical Thermochemistry, 2nd Ed. (1956) Pergamon Press, London.